The New Genetic Medicine

The New Genetic Medicine

Theological and Ethical Reflections

Thomas A. Shannon and
James J. Walter

A SHEED & WARD BOOK

ROWMAN & LITTLEFIELD PUBLISHERS, INC.
Lanham • Boulder • New York • Oxford

A SHEED & WARD BOOK

ROWMAN & LITTLEFIELD PUBLISHERS, INC.

Published in the United States of America
by Rowman & Littlefield Publishers, Inc.
A wholly owned subsidiary of The Rowman & Littlefield Publishing Group, Inc.
4501 Forbes Boulevard, Suite 200, Lanham, Maryland 20706
www.rowmanlittlefield.com

PO Box 317
Oxford
OX2 9RU, UK

British Library Cataloguing in Publication Information Available

Library of Congress Cataloging-in-Publication Data

Shannon, Thomas A. (Thomas Anthony), 1940–
The new genetic medicine : theological and ethical reflections / Thomas A. Shannon and James J. Walter.
p. cm.
Includes bibliographical references and index.
ISBN 0-7425-3170-8 (alk. paper)—ISBN 0-7425-3171-6 (pbk. : alk. paper)
1. Medical genetics—Moral and ethical aspects. 2. Medical genetics—Religious aspects. 3. Biotechnology—Moral and ethical aspects. 4. Biotechnology—Religious aspects. I. Walter, James J. II. Title.

RB155 .S43 2003
174'.296042—dc21 2003008463

Printed in the United States of America

∞™ The paper used in this publication meets the minimum requirements of American National Standard for Information Sciences—Permanence of Paper for Printed Library Materials, ANSI/NISO Z39.48-1992.

Contents

Introduction

Thomas A. Shannon and James J. Walter

In 1953 Francis Crick and James Watson discovered the double helical structure of the DNA (deoxyribonucleic acid) molecule, and this discovery signaled that the classical genetics of Gregor Mendel (+1884), the Austrian monk who had discovered the laws of heredity by working with garden peas, would soon begin to give way to the study and application of the new molecular genetics. In 1953 there was no biotechnology. Yet fifty years later we have mapped and sequenced the human genome, identified many genes responsible for disease, and are initiating gene therapy.

Two important areas that this book will study are the interrelated areas of medical genetics and biotechnology, each of which has had a rapid period of development since the 1970s. Medical genetics is "the aspect of human genetics that is concerned with the relation between heredity and disease."[1] Though the categories of the field overlap, many would consider the following list to comprise medical genetics: diagnosis of genetic disease (e.g., preimplantation diagnosis, prenatal diagnosis, population screening); eugenics as the elimination of disease-related traits (e.g., through selective mating, sterilization); gene therapy on either body cells or germ cells; genetic enhancement of body or germ cells; patenting human gene sequences or techniques of genetic control; embryo research, including pluripotent stem cell research; cloning; and genetic testing on controlled groups.[2]

Biotechnology, on the other hand, involves any technique that uses living organisms to make or modify products, to improve plants or animals, or to develop microorganisms for specific uses.[3] It involves using biology to discover, develop, manufacture, market, and sell products and services. Many of the applications of biotechnology will be used in the health care arena, and this is where the linkage to medical genetics is made. As an industry, biotechnology was first developed in the early 1970s by about ten to twenty venture-backed companies that were involved in pharmaceuticals.[4] Because of these powerful biotechnologies that allowed dramatically increased control over the design of living organisms, controversies over biotechnology and its applications to plants, animals, and humans followed immediately.[5]

What are some of the developments in genetic medicine and biotechnology—along with the ethical reflection that has accompanied them—that have occurred since the 1970s? Surely, recombinant DNA (rDNA) research that was developed in the mid-1970s is a watershed for much of the work that is being done today in both medical genetics and biotechnology. Known as gene splicing, recombinant DNA research is a procedure whereby segments of genetic material from one organism are transferred to another organism or species. The basis of the technique lies in the use of special enzymes (restriction enzymes) that split DNA strands wherever certain sequences of nucleotides occur. These procedures led to the production of vaccines against a number of diseases and the development of certain bioengineered substances such as insulin, interferon, and growth hormone. The pharmaceuticals developed from this technique have substantially improved patient outcome and quality of life, but many have also questioned the wisdom of "playing God" with the very molecules that make up all life.[6] In 1976, the National Institutes of Health established the Recombinant DNA Advisory Committee (RAC) to study and evaluate various clinical proposals to utilize this new scientific technique.

In the last twenty years, scientists have discovered that DNA is virtually interchangeable among animals, plants, bacteria, and humans. Consequently, biotechnology has been extended to plants and animals where the DNA from one animal or plant is spliced into the genetic material of another. There have been mixed reactions to these biotechnologies around the world. In the United States, where people tend to be pragmatic and more willing to take risks with the environment, the biotech companies are moving forward with great speed to produce transgenic plants—but under various federally mandated guidelines. Things may be moving forward even more rapidly in China where, in 1986, Chinese scientists were already aggressively pushing for governmental efforts to bioengineer plants to feed their 1.2 billion people. On the other hand, even though people in the United Kingdom and Germany have staged many protests against the introduction of these genetically altered foods into their countries, the European Union (EU) introduced new laws in March 2003 that will allow foods made from genetically modified organisms into EU countries.

In October 1990 the publicly funded Human Genome Project (HGP) was launched in the United States at an estimated cost of $3 billion. Its aim was to map and sequence all the genetic material of the human person. The project ended approximately five years early in June 2000, and a rough map was unveiled in February 2001. We now know that the human person possesses far fewer genes than previously thought—instead of 80,000 to 100,000 genes we have 30,000 to 40,000 (and maybe as few as 26,000). The knowledge that we have gained from this scientific endeavor and the biotechnologies based on and developed from this map will truly revolutionize clinical medicine over the next twenty to thirty years. Cancer treatments will soon be developed to treat the specific DNA of the patient, personalized pharmaceuticals will be created and marketed by the pharmaceutical industry, and physicians will be able to splice out defective sequences of genes and replace them with proper ones.

Now, the newest scientific venture is concerned with the proteome, or the mapping of all the human proteins coded by genes. Enormous ethical, social, and legal problems are already being raised about the medical genetics and the proteomics that will be developed out of these efforts by big science. This new area of medicine raises some important questions. How will we protect the confidentiality of a person's genetic information? Should we apply the new technology involved in gene therapy to enhancing ourselves? Who owns the rights to the genes and their sequences once they are discovered? Who will have fair access to the new biotechnologies that will be developed? Will we discriminate on the basis of genetic makeup, such that we will deny health insurance or employment to those who possess certain genetic traits or defects? The U.S. Department of Energy has established the Ethical, Legal, and Social Issues Program (ELSI) to study many of these issues, but the questions seem almost endless. Religious groups wonder about the wisdom of "playing God" with DNA,[7] and they are forced many times to rethink their traditional doctrines of creation, divine providence, and redemption.[8]

Another area where medical genetics and biotechnology come together to present us with great potential opportunities but also with tremendous ethical questions are human cloning and human embryonic stem cell research. Once Dolly the sheep was successfully cloned by Ian Wilmut in Scotland, the expectation by some has been that we could or should do the same with humans. Wilmut succeeded by taking a body (somatic) cell that contained all the genetic material to make an exact duplicate of the donor sheep and inserted it into an egg whose own nuclear DNA had been destroyed. Once the full complement of nuclear DNA from the body cell was fused with the egg by using a small electrical charge, the fertilized ovum began to divide and grow. The National Bioethics Advisory Commission (NBAC) was asked to study the scientific and ethical dimensions of human cloning and to make recommendations to the president of the United States and Congress. NBAC recommended against the application of this technology to humans on the basis of its lack of safety.[9]

But one does not have to use this technology to duplicate a whole human person; one could use it with stem cell research to grow organs or tissues for a wide range of diseases, e.g., Alzheimer's disease or Parkinson's disease. Pluripotent stem cells exist as very special cells in the early embryo (ES cells) or in the primordial reproductive cells of the developing fetus (EG cells). Since they have the capacity to become almost any of the 210 cell lines in the mature human body, many believe that the retrieval and differentiation of these special cells could benefit hundreds of thousands of patients. However, to retrieve these cells one must destroy the early embryo or retrieve them from aborted fetuses, and either of these procedures presents great ethical problems for some in our society, especially if the research is funded by public monies. Once again, President Clinton asked NBAC to study the scientific, ethical, and religious issues related to this technology. The commission's report was complex, but it did recommend the public funding of the derivation and use of pluripotent stem cells from both embryos remaining after infertility treatments in *in vitro* fertilization labs and from cadaveric fetal tissue.[10] More recently, in August 2001, President George W. Bush permitted federal funding for research using stem cells from some sixty-three embryonic stem cell lines already in existence. He also established a Council on Bioethics on November 28, 2001, to continue the study of this and other issues such as cloning.

If one did pursue this course, the technology could be combined with human cloning for therapeutic purposes. In this case, the scientist would take a body cell from a patient, extract the nuclear DNA from his or her somatic cell, and insert its nucleus into a human egg whose own nucleus has been destroyed. Once fused, the fertilized egg would be allowed to develop to the four- to six-cell stage and then the pluripotent stem cells would be removed, destroying the embryo in the process. Then the stem cells would be differentiated into the tissue needed by the patient, e.g., spinal cord tissue or brain tissue. Since the tissue would have been cloned from the patient's own body cells, there would no need for special drugs to suppress the patient's immune system after the transplant. Many believe that there is great promise in this new technology when combined with medical genetics, but others question the morality of such procedures that rely on the taking of one life to save another.[11]

As noted, these and other technologies emerging from the HGP are raising substantive religious and ethical questions. In fact, we may be in the biological equivalent of Galileo's demonstration of the heliocentric system. For example, the fact that DNA is common to all living organisms and, more specifically, that human DNA differs from the DNA of orangutans by 0.2 percent and that we differ from the mouse genome by 300 genes raises profound questions about the definition of human nature and what, if anything, might give us predominance over other creatures.

The purpose of bringing together this collection of essays on theological and ethical issues related to genetic medicine is to begin to examine these questions, as well as their implications. Of particular importance for this project is our subtitle: *Theological and Ethical Reflections*. We are proposing these essays as a kind of first draft examination of the questions. We do not consider our essays as a final draft or even where we might find ourselves in several years. Part of our tentativeness is the rapidly changing nature of the genetic information itself. This is a field of study that is under continuous construction. New information and possible applications come at us each day. One can hardly digest the latest discovery before the new one is at hand. Additionally, the complexity of the information is staggering. Even those in the field have difficulty understanding what their colleagues are doing because of the specialized nature of the research. But it is important that the difficult and complex work of beginning this examination begin. Yet we need to remember that work in this area will always be a work in progress.

Furthermore, we write from a self-consciously Roman Catholic perspective. The operative word here is *perspective*. We stand *within* the tradition but do not understand that our task as theologians and ethicists is simply to repeat the tradition, and a selected part at that. Rather, we understand our responsibility first to understand the tradition as best as we can. This aspect is most critical, for what is significant in the Catholic tradition is an understanding of how this tradition has grown, developed, and changed over the centuries. Second, we see our task as bringing this history to bear on the precise questions posed by contemporary genetics. While it is true that many of the thematic values inherent in the Catholic tradition are relevant to the questions at hand, we are in a radically new situation with new capacities and powers simply unimagined by earlier formulations of this tradition. We need to start thinking about these problems. In addition, we recognize that some of our discussions and positions might not cohere with some magisterial teachings. On the one hand, we want to affirm the role of the magisterium in the tradition but, on the other hand, repeating traditional formulae does not always seem to address in a substantive way the precise issues under scrutiny. While we hope our attitude and reasoning are respectful of the magisterium, we find it necessary to be much more interdisciplinary in our reasoning and open to other methodologies, some from within the tradition and some from other perspectives.

A final word about the structure of this book. We begin with some initial theological and ethical reflections on genetics in general and the HGP in particular. From there, we focus our analyses on some of the more specific issues in contemporary genetic medicine, e.g., the definition of human nature in light of modern genetics, human gene transfer, the moral status of the preimplantation embryo, embryonic pluripotent stem cell research, and human cloning. Finally, we end with a series of theological and ethical reflections on the bioengineering of planet Earth: the remaking of plants, animals, and even human beings in light of the new scientific discoveries discussed earlier in the volume.

NOTES

1. Victor A. McKusick, *Human Genetics*, 2d ed. (Englewood Cliffs, N.J.: Prentice-Hall, 1969), 181.
2. Lisa Sowle Cahill, "Genetics, Ethics and Social Policy: The State of the Question," in *Concilium: The Ethics of Genetic Engineering*, edited by Maureen Junker-Kenny and Lisa Sowle Cahill (London: SCM, 1998), vii.
3. Nanette Newell, "Biotechnology," in *Encyclopedia of Bioethics*, rev. ed., vol. 1, edited by Warren T. Reich (New York: Simon and Schuster Macmillan, 1995), 283.
4. Alison Taunton-Rigby, *Bioethics: The New Frontier* (Bentley College, Mass.: Center for Business Ethics, 2000), 6–8.
5. For the authoritative encyclopedia on this topic, see the *Encyclopedia of Ethical, Legal, and Policy Issues in Biotechnology*, 2 vols., edited by Thomas H. Murray and Maxwell J. Mehlman (New York: John Wiley and Sons, 2000).
6. For example, see the President's Commission for the Study of Ethical Problems in Medicine and Biomedical and Behavioral Research, *Splicing Life: The Social and Ethical Issues of Genetic Engineering with Human Beings* (Washington, D.C.: U.S. Government Printing Office, 1982).
7. For a helpful essay on the meaning of the phrase "playing God," see Allen Verhey, "'Playing God' and Invoking a Perspective," *The Journal of Medicine and Philosophy* 20 (1995): 347–64.
8. For example, see Ronald Cole-Turner, *The New Genesis: Theology and the Genetic Revolution* (Louisville: Westminster/John Knox, 1993).
9. National Bioethics Advisory Board, *Cloning Human Beings*, 2 vols. (Rockville, Md.: U.S. Government Printing Office, 1997).
10. National Bioethics Advisory Board, *Ethical Issues in Human Stem Cell Research*, 3 vols. (Rockville, Md.: U.S. Government Printing Office, 2000).
11. See the President's Council on Bioethics Report, *Human Cloning and Human Dignity: An Ethical Inquiry*, http://www.bioethics.gov/report.html, July 2002.

1

Theological Issues in Genetics

James J. Walter

Contemporary molecular genetics and reproductive medicine are posing far-reaching questions for theological reflection. This chapter will focus on three topics that bring to light some of these questions: (1) human gene transfer, (2) somatic cell nuclear transplant cloning of humans, and (3) patenting of human genes. Two theological frameworks or hermeneutical themes that are currently shaping and informing the moral debates about these topics will be selected: first, the human as created in the image of God (*imago dei*) and, second, the human as "playing God." Both have biblical foundations and are frequently used to describe the human person created good in the divine image, but, since the Fall, prone to hubris and the irresponsible exercise of freedom. My analysis will focus on statements from Christian councils and task forces, as well as on various pronouncements from ecclesiastical communities and theological writings on these topics.

IMAGO DEI

The distinctiveness of humans in the plan of creation is often described by reference to the fact that humans are created in the image of the divine. However, given the variety of meanings of *imago dei* within Christian theology, only two will be discussed: stewardship and created cocreatorship.[1] Deciding which of these interpretations to select relies not only on how one reads scripture (especially Genesis, Psalm 8, and Gospels accounts of Jesus's healing of the sick) but also partially on where one stands vis-à-vis two important theological themes: (1) the nature and extent of human responsibility to pursue genetic progress, and (2) the theological doctrine that grounds both human intervention into genetic material and our knowledge of God's purposes.

Human Gene Transfer

There are four types of human gene transfer that are likely to be developed as a result of the Human Genome Project: (1) somatic cell therapy, (2) germ-line therapy, (3) somatic cell enhancement, and (4) germ-line enhancement. Nearly all the task forces, ecclesiastical communities, and theologians who have addressed the first of these types have approved of its use once the scientific and technical difficulties have been solved.[2] On the other hand, many of the same have rejected the use of both types of enhancement gene transfer (somatic and germ-line).[3] The one remaining form of human gene transfer, germ-line therapy, remains theologically and morally the most contentious.[4]

Examination of the positions on human gene transfer clearly reveals different theological models of the *imago dei* that shape and inform the authors' moral visions and judgments. Stewardship over creation is historically one of the most frequently used models of the *imago,* and it accentuates the fact that humans are entrusted with responsibility for conserving and preserving creation. It tends to place limits on human freedom to alter what the divine has created, and sometimes it claims some knowledge of God's purposes by reference to a doctrine of creation. The National Council of Churches adopts many aspects of this theological model as the presupposition of its moral acceptance of somatic cell therapy. The human role in the creative process of creation is one of being responsible for what the divine has made and living in harmony with all creation.[5] Joseph Cassidy and Edmund Pellegrino also argue morally for somatic cell therapy and against all forms of enhancement technologies based on their view of the human as the steward (not cocreator) over human germ plasm for future generations. In addition to the sources of scripture and the teachings of the magisterium, they contend we have an insight into God's will by using the knowledge God has built into creation.[6]

Another aspect of the *imago* has recently emerged in both Protestant and Catholic circles, and it is most often characterized as "created co-creator."[7] This model recognizes that we are indeed created beings, and thus we ultimately rely on the divine for our existence. Though only God creates *ex nihilo,* we do mirror the divine in our capacity to create, even if that ability is restricted to fashioning materials already in the created order. Since creation itself is not complete (*creatio continua*), we have responsibilities to help bring it to completion. Furthermore, because we cocreate with the divine, we have greater freedom than in the previous model to intervene into our genetic material. Ronald Cole-Turner has adopted this model in his moral acceptance of somatic cell therapy. He argues that, though the divine works through the processes of nature, God's creative intentions transcend nature. As cocreators, we must discover these divine purposes so that we might intervene into the *moral* disorder within nature, i.e., the disorder that is both pervasive and an inevitable byproduct of the evolutionary process, to correct it. Cole-Turner does not turn to the doctrine of creation for knowledge of God's will but to the doctrine of redemption, which provides the necessary noetic clue to God's purposes for curing genetic diseases.[8]

Ted Peters takes this model one step further by morally justifying not only human gene transfer for therapeutic purposes but also for the ends of enhancement (somatic and germ-line). He grounds human responsibility and knowledge of God's purposes neither in the doctrine of creation nor in redemption but in the doctrine of eschatology. He argues that the created cocreator model is superior to all others because it begins with a vision of openness to God's future and responsibility for the human future. Such a vision is founded on our vision of the promised Kingdom of God, and this framework of future possibilities orients and directs our moral activity in genetics.[9] Prolepsis is the structure of ethical reasoning, and it is a concrete actualization within the present of what we see to be the case in the future-transformed reality. For Peters, we must begin our ethical thinking about human gene transfer by projecting a vision of the new creation and then work back to the present to discover our moral responsibilities.[10]

Somatic Cell Nuclear Transplant Cloning

Applying the same framework to human cloning indicates again that moral judgments are shaped by theological models of the *imago dei*. Because this type of cloning will be employed in the future as a form of assisted reproductive technology, the theological theme of responsibility for intervening into and improving on human reproduction becomes relevant.

A large portion of the literature does not support morally this type of human cloning, and most utilize theological resources to inform their positions.[11] Nevertheless, there is a sizable minority that either has cautiously endorsed human cloning or at least can find no convincing theological reason to oppose it categorically.[12] Those who use the stewardship model of the *imago* to frame this issue *tend* not to support it morally, and those who use the created cocreator model *tend* to permit, or at least not to oppose, it on theological grounds. Two examples illustrate this point.

The United Methodist Genetic Science Task Force's statement on cloning begins with a theological affirmation of how humanity is created in the image of God as stewards, and then it develops its policy statement calling for a ban on human cloning.[13] On the other hand, Ted Peters, who has been an ardent proponent of the cocreator model, has strongly contended, "Reproductive and genetic technologies, along with technologies to create a child through cloning, can express responsible created cocreatorship."[14]

Patenting Human Genes

The scientific laboratories that are mapping and sequencing the human genome are also applying for patents on the genes they discover. There are numerous moral and legal problems with this topic, but there are many theological issues as well. One such issue involves a theological discussion of whether or not the *imago dei* is present in human DNA and thus that genes deserve to be treated differently from other created material.

Richard Land and C. Ben Mitchell argue morally against the patenting of human genes, and part of their argument is based on a theological construal of the *imago dei.* They assert that only human beings are created in the image of the divine, and this image "pervades human life in all of its parts."[15] Others have argued that DNA provides the biological blueprint for humans as the image of God, and thus all patenting of human genetic material is inappropriate.[16] On the other hand, Ted Peters does not accord human DNA any special status and argues that it is precisely because we are created in the image of God (created cocreators) that we are called to use our creativity to make this world a better place. If the patenting of DNA can further this cause by eliminating debilitating genetic diseases, then on the basis of this image of God we should morally permit the patenting of life.[17]

PLAYING GOD

A second theological framework that has shaped the discussion of these contemporary topics in genetics concerns the question of whether or not humans are exceeding their limits, and thus playing God, by intervening into the very material that constitutes life. Of course, where one stands on this question is partially determined by which theological model of the *imago dei* one adopts. Because the stewardship model tends to limit human activity through its emphases on conserving and preserving creation, the claim of improperly playing God will frequently arise from those who subscribe to this model. The reverse tends to be the case for those who argue for a created cocreator model. Beyond that, however, this framework implies two theological themes: (1) the status of human DNA, and (2) one's position on the sovereignty of God and the divine ownership of creation.

Though there has been a spate of books that include this phrase somewhere in their titles,[18] what is clear is that there is no common understanding of what "playing God" means.[19] Some find the phrase not very helpful and believe that bioethical discussions could be enhanced without its use,[20] while others argue that it can serve as an important and distinctively theological perspective from which to assess scientific and technological innovations. When used as a theological perspective it can function either negatively as a concern or warning, with specific prohibitions attached to it, or positively as an invitation to "play God" by imitating God's purposes of care and grace. Consequently, as a perspective there can be improper and proper forms of playing God.[21]

Human Gene Transfer

When the president's commission on genetic engineering submitted its report in 1982, it noted that there was an objection from religious groups that scientists were playing God in their recombinant DNA (rDNA) research. When the phrase is applied to this research, which is the precursor to techniques in human gene transfer, the commission interpreted "playing God" as a concern about the consequences of exercising great human powers over nature.[22] The National Council of Churches echoed a similar definition when it used this phrase to describe the fact that human beings now possess the ability to do "God-like" things, i.e., to direct and redirect the life processes of nature itself.[23] The Church of the Brethren took this concern one step further and questioned whether humans are now playing God by changing the genetic structures of life and thus overstepping the boundaries God has set for humankind.[24] Each of these views, then, raises a concern about human intervention into DNA, whether at the somatic or germ-line level. Part of this concern is over the status of DNA or the human genome: is it sacred and thus beyond the boundaries of human manipulation and control, or is it more or less similar in status to other bodily matter and thus open to human intervention and control?[25] If the former, then scientists are improperly playing God when they intervene into the human genome because they overstep the boundaries given to them by God; if the latter, then scientists are properly playing God, i.e., serving God's own purposes, when they intervene at some level (somatic or germ-line) to cure and prevent disease or, for some, to enhance the human. Of the literature surveyed, it appears that most authors have argued that, though human DNA is a cause of great wonder, it does not possess a sacred status but is like other parts of the body and thus in principle may be altered within certain therapeutic limits.[26] It goes without saying that those few who have supported morally somatic or germ-line enhancements have denied the special or sacred status of DNA.[27]

Somatic Cell Nuclear Transplant Cloning

Lee Silver, a biologist, contends that all ethical arguments used to prohibit human cloning are really hidden religious arguments about the wrongfulness of playing God.[28] As intriguing as this claim might be, in the end it misunderstands much of the contemporary theological debate[29] and the important disputes that took place between Joseph Fletcher and Paul Ramsey in the 1960s and 1970s on the topics of human cloning and assisted reproductive technologies. Two of the most often-quoted phrases about playing God were constructed during these decades and on this topic. Fletcher's bold claim "Let's play God" laid down the gauntlet for those who would morally oppose the development of these new technologies.[30] Ramsey's famous phrase sounded an alarm of caution to those who would arrogate to themselves control over human reproduction: "Men ought not to play God before they learn to be men, and after they have learned to be men they will not play God."[31] Fletcher's perspective on "playing God" had little to do with God; in fact, he used this phrase as a way of signaling the death of the "God of the Gaps" and the need for humans to take up their responsibilities.[32] These responsibilities, however, were fashioned out of a Baconian desire to dominate and control the processes of human reproduction, thus the subtitle of his book. Ramsey's perspective on playing God vis-à-vis cloning was much different; it was a warning that, whenever God is absent or superfluous, humanity becomes the Creator and engineer of the future and nature, and human nature will be controlled with messianic ambition.[33] The results of this situation for Ramsey were that morality would be reduced to consequentialism and human nature would be left with no dignity of its own.[34]

Patenting Human Genes

When the framework of playing God is used to inform moral deliberation and judgments about the patenting of human genes, once again we find the theological themes of the status of human DNA and the sovereignty of God over creation. In general, those who have invoked this perspective of playing God seek to show how this form of patenting is tampering with the blueprint for life forms and the arrogant disregard for God's ownership over life.[35] Thus, patenting human genes is considered an improper form of playing God.

Some have argued theologically that human DNA and the genome itself are sacred because they possess characteristics integral to human identity and personhood. Furthermore, DNA provides the biological blueprint for humans created in the image of God, and it is even possible to accord this genetic material the social and cultural functions of the soul.[36] In addition, Land and Mitchell argue that the very image of God in humans pervades human life in all its parts, and this certainly includes DNA.[37] Several argue that to patent human genes is to play God improperly because such actions take away God's sovereign ownership over these genetic materials. For example, the Southern Baptist Convention and the United Methodist community both have argued against such patenting on the theological ground that only God can and does own life.[38]

On the other hand, those who are open to some patenting of human genes have not used at all the framework of playing God, and they have not argued for the sacred status of human DNA. For example, Cole-Turner contends that theists believe only God is sacred, and thus everything else is God's creation. He argues there is no metaphysical difference between DNA and other complex chemicals, and so there is no distinctly religious grounds for objecting to the patenting of DNA.[39] Finally, Mark Hanson has argued that there are two significant problems with the divine ownership claim: first, patents do not confer ownership, so the theological claim against human ownership does not hold; and second, such arguments based on divine sovereignty are consistent with a narrow conception of some doctrines of God.[40] He contends that other views of God's ownership need to be constructed, e.g., one that might approach divine ownership as God's reserving the right to define the purpose and value of created realities.[41] A brief summary of the key theological issues on the selected bioethical topics will serve as a conclusion to this section. It is important to note that theological reflection can contribute a distinct perspective from which to evaluate morally these complex topics. Theological frameworks or hermeneutical themes can be constructed from the Christian faith that in turn possess the power to shape and inform moral assessment. For the moral judgments on these selected topics rely, at least partially, on prior judgments about (1) the nature and extent of human responsibility to pursue genetic progress; (2) the limits, if any, to alter the very material that constitutes life; and (3) the moral status of human DNA. Theological construals of God's purposes in creation, of the sovereignty of God over creation, and of divine ownership of human life are key to arriving at judgments on these questions.

Review of various ecclesiastical statements and of theological opinions has indicated that there is no universal agreement on the morality of these topics. One suggestion that might be offered for future discussion concerns the importance of focusing reflection more squarely on the theological presuppositions to these bioethical topics. The development of a consensus on these central theological issues at stake might prove helpful as a first step in articulating an acceptable range of moral judgments.

NOTES

1. Two other models have been recently proposed as well. James Gustafson has proposed a "participant" model in his *Ethics from a Theocentric Perspective, Volume Two: Ethics and Theology* (Chicago, Ill.: University of Chicago, 1984), 13, 294; and Arthur Peacocke has proposed a "co-explorer" model in his *Creation and the World of Science* (Oxford: Clarendon, 1979), 304–06.

2. For example, see John Paul II, "Biological Research and Human Dignity," *Origins* 12 (October 22, 1982): 342–43; idem, "The Ethics of Genetic Manipulation," *Origins* 13 (November 17, 1983): 385, 387–89; Science and Human Values Committee of the NCCB, "Critical Decision: Genetic Testing and Its Implications," *Origins* 25 (May 2, 1996): 769, 771–72; The [British] Catholic Bishops' Joint Committee on Bioethical Issues, *Genetic Intervention on Human Subjects: The Report of a Working Party of The Catholic Bishops' Joint Committee on Bioethical Issues* (London: The Linacre Centre, 1996), 28, 42; World Council of Churches, *Manipulating Life: Ethical Issues in Genetic Engineering* (Geneva: Church and Society, 1982), 6; National Council of the Churches of Christ in the U.S.A., *Human Life and the New Genetics: A Report of a Task Force Commissioned by the NCC* (New York: Office of Family Ministries and Human Sexuality, 1980), 43; Seventieth General Convention of the Episcopal Church (July 1991); United Church of Christ, "The Church and Genetic Engineering" (Seventeenth General Synod of the UCC, Forth Worth, Tex., June 29–July 4, 1989), 1–5, at 3; and The United Methodist Church Genetic Science Task Force Report to the 1992 General Conference, "New Developments in Genetic Science Challenge Church and Society," *Church and Society* (1992), 113–23, at 121. Also, see Pius XII's 1953 address on genetics, "Moral Aspects of Genetics," in *Medical Ethics: Sources of Catholic Teachings*, 2d ed., edited by Kevin D. O'Rourke, O.P., and Philip Boyle (Washington, D.C.: Georgetown University, 1993), 130–31.

3. For example, see Paul Abrecht, editor, *Faith and Science in an Unjust World: Report of the World Council of Churches' Conference on Faith, Science and the Future, Volume 2: Reports and Recommendations* (Philadelphia: Fortress, 1980), 66; United Methodist Church, "New Developments in Genetic Science," 122; Allen Verhey, "'Playing God' and Invoking a Perspective," *The Journal of Medicine and Philosophy* 20 (1995): 347–64, at 361; and the summary found in Ted Peters, "Intellectual Property and Human Dignity," in *The Genetic Frontier: Ethics, Law, and Policy*, edited by Mark S. Frankel and Albert Teich (Washington, D.C.: American Association for the Advancement of Science, 1994), 215–24, at 221. Interestingly, John Paul II and the Catholic Health Association in the U.S. have not taken, in principle, negative positions on either form of these genetic enhancements. See John Paul II, "The Ethics of Genetic Manipulation," 385, 387–89, at 388; and The Catholic Health Association of the United States, *Human Genetics: Ethical Issues in Genetic Testing, Counseling, and Therapy* (St. Louis, Mo.: CHA, 1990), 22. Also, Ted Peters has argued that we should not close the door to enhancement technology. See his "'Playing God' and Germline Intervention," *The Journal of Medicine and Philosophy* 20 (August 1995): 365–86, at 365.

4. For a summary of the moral and theological arguments for and against this form of therapy, see James J. Walter, "'Playing God' or Properly Exercising Human Responsibility? Some Theological Reflections on Human Germ-Line Therapy," *New Theology Review* 10 (November 1997): 39–59.

5. National Council of Churches of Christ, *Human Life and the New Genetics*, 42. Also, Panel on Bioethical Concerns of the NCC/USA, *Genetic Engineering: Social and Ethical Consequences* (New York: Pilgrim, 1984), 24.

6. Joseph D. Cassidy, O.P., and Edmund D. Pellegrino, "A Catholic Perspective on Human Gene Therapy," *International Journal of Bioethics* 4 (1993), 11–18, at 12. Another way to ground our moral agency and responsibilities is in the doctrine of the incarnation. See John S. Feinberg and Paul D. Feinberg, *Ethics for a Brave New World* (Wheaton, Ill.: Crossway, 1993), 280.

7. The Protestant theologian Philip Hefner is generally credited with the naming of this model, but the Jesuit theologian Karl Rahner had already anticipated the substance of the model in the late 1960s. See Philip Hefner, "The Evolution of the Created Co-Creator," in *Cosmos as Creation: Theology and Science in Consonance*, edited by Ted Peters (Nashville: Abingdon, 1989), 211–33. For Rahner's two widely read and influential essays, see "The Experiment with Man: Theological Observations on Man's Self-Manipulation," in *Theological Investigations*, vol. IX, trans. Graham Harrison (New York: Herder and Herder, 1972), 205–24; and "The Problem of Genetic Manipulation," ibid., 225–52. For an interesting comparison between Rahner's two articles, see David F. Kelly, "Karl Rahner and Genetic Engineering: The Use of Theological Principles in Moral Analysis," *Philosophy and Theology* 9 (Autumn–Winter 1995): 177–200.

8. Ronald Cole-Turner, "Is Genetic Engineering Co-Creation?" *Theology Today* 44 (October 1987): 338–49, at 345–47; and idem, *The New Genesis: Theology and the Genetic Revolution* (Louisville, Ky.: Westminster/John Knox, 1993), 98–103.

9. Ted Peters, *Playing God? Genetic Determinism and Human Freedom* (New York: Routledge, 1997), 144–56.

10. Ted Peters, *For the Love of Children: Genetic Technology and the Future of the Family* (Louisville: Westminster/John Knox, 1996), 155.

11. For example, the Congregation for the Doctrine of the Faith, "Instruction on Respect for Human Life in Its Origin and on the Dignity of Procreation," *Origins* 16 (March 19, 1987). 1, at paragraph 6, Cardinal Keeler, "The Problem with Human Cloning," *Origins* 27 (February 26, 1998): 597 and 599–601; General Assembly of the Church of Scotland, "Motions on Cloning," in *Human Cloning: Religious Responses*, edited by Ronald Cole-Turner (Louisville, Ky.: Westminster/John Knox, 1997), 138; and The Christian Life Commission of the Southern Baptist Convention, "Against Human Cloning: March 6, 1997," in *Cloning Human Beings: Report and Recommendations of the National Bioethics Advisory Commission*, National Bioethics Advisory Commission (NBAC) (Rockville, Md.: U.S. Government Printing, 1997), 56.

12. For example, see Joseph Fletcher, "Ethical Aspects of Genetic Controls," *New England Journal of Medicine* 285 (April 1971): 776–83; Ted Peters, "Cloning Shock: A Theological Reaction," in *Human Cloning*, Cole-Turner, 12–24; and Philip Hefner, "Cloning as Quintessential Human Act," *Insights* (August 1997): 18–21.

13. Genetic Science Task Force, "Statement from the United Methodist Genetic Science Task Force: May 9, 1997," in *Human Cloning*, Cole-Turner, 143–45, at 143. Also, see R. Albert Mohler, Jr., "The Brave New World of Cloning: A Christian Worldview Perspective," in *Human Cloning*, Cole-Turner, 91–105.

14. NBAC, *Cloning Human Beings*, 47. Also, see Cole-Turner, who himself adopts a created cocreator model, in which he claims that he does not believe "that

a compelling theological argument can be made against cloning for reproductive or for experimental purposes." Ronald Cole-Turner, "At the Beginning," in *Human Cloning*, Cole-Turner, 119–30, at 120.

15. Richard D. Land and C. Ben Mitchell, "Patenting Life: No," *First Things* 63 (May 1996): 20–22, at 21.

16. See the discussion of this view by Mark J. Hanson, "Religious Voices in Biotechnology: The Case of Gene Patenting," *Hastings Center Report* 27 (November–December 1997): 1–21, at 4.

17. Ted Peters, "Patenting Life: Yes," *First Things* 63 (May 1996): 18–20, at 19. Also, see his *Playing God?* 139.

18. For example, Ted Howard and Jeremy Rifkin, *Who Should Play God? The Artificial Creation of Life and What It Means for the Future of the Human Race* (New York: Delacorte, 1977); and R. C. Sproul Jr., *Playing God: Dissecting Biomedical Ethics and Manipulating the Body* (Grand Rapids, Mich.: Baker, 1997).

19. Some recent attempts have been made to clarify the meaning of the phrase. For example, see Verhey, "'Playing God' and Invoking a Perspective," 347–64; and Lisa Sowle Cahill, "'Playing God': Religious Symbols in Public Places," *The Journal of Medicine and Philosophy* 20 (1995): 341–46.

20. See Howard Brody, *Ethical Decision in Medicine* (Boston: Little, Brown and Co., 1976), 82.

21. Verhey, "'Playing God' and Invoking a Perspective," 358.

22. President's Commission for the Study of Ethical Problems in Medicine and Biomedical and Behavioral Research, *Splicing Life* (Washington, D.C.: Government Printing Office, 1982), 54.

23. NCC, *Genetic Engineering*, 27.

24. Church of the Brethren, "1987 Annual Conference Statement on Genetic Engineering" (1987 Annual Conference Minutes), 451–56, at 453.

25. For a discussion of some of these issues, see Bernard Baertschi, "Devons-Nous Respecter Le Génome Humain?" *Revue de Théologie et de Philosophie* 123 (1991): 411–34.

26. See Cassidy and Pellegrino, "A Catholic Perspective on Human Gene Therapy," 12; and Report of the Working Party of the [British] Catholic Bishops, *Genetic Intervention on Human Subjects*, 32. Jeremy Rifkin, who has protested against this view, is a notable exception. See his *Algeny* (New York: Penguin, 1983).

27. Peters, *Playing God*, 117.

28. Lee M. Silver, "Cloning, Ethics, and Religion," *Cambridge Quarterly of Healthcare Ethics* 7 (1998): 168–72, at 169.

29. For example, see the argument against human cloning based on biological diversity in Richard A. McCormick, S.J., "Should We Clone Humans?" *Christian Century* 110 (November 17–24, 1993): 1148–49.

30. Joseph Fletcher, *The Ethics of Genetic Control: Ending Reproductive Roulette* (Garden City, N.Y.: Anchor, 1974), 126.

31. Paul Ramsey, *Fabricated Man: The Ethics of Genetic Control* (New Haven, Conn.: Yale University, 1970), 138.

32. Fletcher, *The Ethics of Genetic Control*, 200.

33. Ramsey, *Fabricated Man*, 91–96.

34. See Verhey, "'Playing God' and Invoking a Perspective," 356.

35. See Rebecca S. Eisenberg, "Patenting Organisms," in *Encyclopedia of Bioethics*, rev. ed., vol. 4, edited by Warren T. Reich (New York: Simon and Schuster Macmillan, 1995), 1911–14, at 1911.

36. For a summary of this position, see Hanson, "Religious Voices in Biotechnology," 4.

37. Land and Mitchell, "Patenting Life: No," 21.

38. Report of Committee on Resolutions, "On the Patenting of Animal and Human Genes," *SBC Bulletin* (1995): 7–8, at 7; and United Methodist Church Task Force, "New Developments in Genetic Science," 117.

39. Ronald Cole-Turner, "Religion and Gene Patenting," *Science* 270 (October 6, 1995): 52.

40. Hanson, "Religious Voices in Biotechnology," 9–10.

41. I would like to acknowledge the generous assistance that I received from my graduate assistant Timothy Sever and from John H. Evans of Princeton University in preparing this manuscript.

2

Ethical Issues in Genetics

Thomas A. Shannon

PRIVACY

Three types of privacy have been identified: physical (freedom from physical contact), informational (which limits access to information about one's self), and decisional (the capacity to make decisions for one's self).[1] All are impacted by both genetic testing and various forms of prenatal diagnosis.[2] While the means of diagnosis are minimally invasive physically (a drop of blood or a single strand of hair is enough), obtaining such samples can constitute invasions of physical privacy. To learn whether there is a genetic component to a disease, elaborate family pedigrees must be constructed. Knowledge that one family member has a genetic predisposition for a disease has implications for other family members. If screening is a precondition for either insurance or employment, substantive privacy and social issues are raised. While insurance companies already screen potential customers through physical exams, some fear that genetic screening will also be required, either through direct testing or the disclosing of previously taken tests.[3] Thus individuals with a genetic disposition for breast cancer may be uninsurable because of the financial loss they represent to a company. Employers too have an interest in learning the genetic profiles of present or potential employees, but access to such information can violate both informational and decisional privacy.[4] While issues of discrimination and paternalism can arise in companies' employment policies, nonetheless some employees could be at risk because of certain jobs. While rectifying the environment is certainly one way to help resolve this, individuals still remain sensitive to certain pollutants. Public policy issues for these agenda have not been resolved. Because genetic information is both private and social, we are only beginning to realize the impact that genetic screening will have on our traditional understanding of privacy and confidentiality.

Similar issues arise in prenatal diagnosis[5] that makes the health status of the fetus immediately accessible and visible and thus available to insurance companies. This potentially compromises both the confidentiality of such information and the mother's decisional privacy by limiting her range of options, especially if the insurance company determines that the condition of the fetus is a preexisting one and will provide no reimbursement for medical care.

RISK-BENEFIT ANALYSIS

Risk-benefit analysis is a traditional way of deciding whether or not to undergo a particular procedure. While offering the promise of many benefits, new genetic interventions also present a new range of risks.[6] A new issue is learning that one may be susceptible to a disease that will occur later in life, such as breast cancer or Huntington's disease—thus the term "presymptomatic" disease.[7]

Because no therapies are yet available other than perhaps a drastic treatment such as prophylactic mastectomy and oophorectomy, some question the value of such information.[8] A particular problem is identifying the predisposition for a disease with the actual disease, which causes additional suffering for the individual, as well as possibly disqualifying them for insurance. Another problematic category is the screening of young children and adolescents. The issue of informed consent is particularly difficult, especially when a minor is becoming mature but nothing can be done to treat the disease. Issues of self-esteem, stigmatization, and complex familial relations are of concern, to say nothing of the previously discussed insurance difficulties.[9]

Another dimension of the risk-benefit problems related to genetic testing (and screening) is the accuracy of the test itself as well as the number of false positives and false negatives it produces. While these are primarily technical questions related to the test itself, they also raise profound ethical questions: When should a test be made available? Is the test applied to all of the possible genes associated with a disease or only the more common sites tested (e.g., with cystic fibrosis, there are over 300 mutations which can cause cystic fibrosis, but typically about 60 to 70 sites are tested)? How expensive will the test be? Will the number of false positives or false negatives cause more harm than not making the test available?

Two important facts must be kept in mind when evaluating or using any genetic testing technologies. First, while literally thousands of genetic anomalies can be detected, we understand the health implications of only a few of them. Second, we cannot cure any of the genetic anomalies that we detect. These two hard realities frame any ethical discussion, particularly discussions of late-onset genetic diseases such as breast cancer. While some interventions can be made that alleviate some symptoms, or the information can be used to prepare families for what is to come, precious little can be done about the disease itself. Thus the choices are poor: either to avoid reproduction, to use donor sperm or egg, to abort, or to continue the pregnancy with the disease running its natural course. If the last option is chosen, little insurance and few social resources will be available to care for the child.

While prenatal diagnosis is offered typically to women in higher risk categories (women with a history of genetic disease or over 35 years of age[10]) and while only about two percent of such diagnoses lead to potential abortion, nonetheless prenatal diagnosis will become more common. First, as new genetic discoveries are announced, pressure will increase to detect these as early as possible. Second, given the current malpractice climate, prenatal diagnosis becomes a means of defensive medicine. Third, as childbearing is being delayed until later in life—with infertility increasing—and as people are having fewer children, pressure builds to have as healthy a child as possible.

Although one's child might have a so-called normal genome, that in itself does not mean that the child will be healthy, never contract a fatal disease, or have a pleasant personality. Genetic screening can raise expectations that cannot be met and unwittingly open the door to a new kind of eugenics, family eugenics. In this case the couple selects a genetic profile in the expectation of obtaining a certain type of child. Since currently one can already order somewhat custom-designed embryos, this application is not far fetched.[11] Prenatal diagnosis may be setting up a situation in which children are desired for specific characteristics, not for who they are.

All of these issues surrounding prenatal diagnosis raise this critical question: What is the problem that prenatal diagnosis is meant to solve?

On the benefit side of the equation is the developing use of human gene transfer.[12] The first intervention is somatic cell therapy, which has three forms: (1) *ex vivo,* in which cells are removed from the body, corrected, and then returned so that the new function can be expressed and correct the disease; (2) *in situ,* in which the new gene is directly introduced into the locus of the disease; and (3) *in vivo,* in which the therapeutic gene is injected into the bloodstream and travels to the proper tissue.[13] The second intervention is germ-line therapy, which corrects an anomaly by placing the corrected copy in the germ cells in the fertilized egg; this both corrects the condition for the individual and also allows the correct copy to be passed on to one's descendent.

In general, the ethical analysis of somatic cell gene therapy follows in broad outline an analysis similar to that of the introduction of any new medical therapy. Walters and Palmer identify seven key questions: (1) What disease is to be treated? (2) Are there alternative forms of therapy, and are they affordable? (3) What is the anticipated or potential harm of the therapy? For example, will the virus used to transport the new genetic material become reactivated and cause harm, will the new genetic material reach the correct part of the cell, will there be any harmful long-term effects? (4) What is the expected or anticipated benefit? (5) Will patients be selected fairly? Children had traditionally been protected by not being included in research projects, but current thinking is that no group should be excluded from research, particularly if gene therapy can be potentially more beneficial when introduced earlier. (6) How will informed consent be ensured? While this question is critical for the adult population who may be desperate for a potentially lifesaving therapy, it is also critical for children whose parents may frantically desire to save their children. (7) How will privacy and confidentially be preserved? Given the highly experimental nature of this research, its inherent newsworthiness, as well as the penchant for feeding frenzies on the part of the media, such concerns are not academic. Yet the identities of the two children who were the first subjects of gene therapy were kept confidential for over a year and eventually released only with the parents' permission.[14]

Three key ethical questions are: (1) How quickly should gene therapy move to clinical practice? Should a particular therapy prove successful, there will be tremendous pressure to move it from the laboratory to the bedside as soon as possible. But we need to remember that the critical ethical variable here is that the therapy must be proven to work and to have at least no negative short-term side effects. (2) How efficacious or successful is the therapy? From 1990 to 1995, 100 clinical trials of gene therapy were initiated. Yet Leiden's assessment of these trials is that "[t]o date, there is little or no published evidence of the clinical efficacy of gene therapy."[15] Leiden does not see this as a condemnation of the field. Rather, he draws three conclusions: that gene therapy is grounded in solid scientific principles, that the negative results so far are a function of the newness of the field, and that recent progress promises optimism for the future. (3) How will this resource be allocated?[16] While the consequences of genetic diseases are severe, the numbers of those affected by a particular genetic disease are relatively small—perhaps between 10,000 and 15,000. Can the cost of research and clinical trials for these diseases be justified? While it is true that much can be transferred to other technologies and strategies, it is even more true that victories will come at a high cost and the other health needs of the nation are increasing. Thus the issues of allocation and priorities need substantive national debate.

Germ-line gene therapy, by both preventing disease by inserting correct copies of genes into reproductive cells and enabling this correction to be passed on to succeeding generations, presents both technical and ethical problems. Wivel and Walters identify four technical problems that need to be resolved before any human trials can be initiated: the inserted gene will need to function normally; the insertion of the new gene must not cause impairment of normal gene function; there must be no residual effects from the original genetic defect; and there must be no genetic side-effects from the insertion of the new gene. Common to these problems is the challenge of physical placement of a new gene in the proper location. But it is also important that the new copy not cause a problem with the other genes near the site of insertion. The interaction of genes with their neighboring genes at locations along various strands is not well understood and is a major scientific obstacle to initiating human trials.

Because the genetic correction will be passed on to one's descendants, germ-line therapy is surrounded by a major debate. The major arguments in favor are that only this type of therapy, precisely because it is initiated on the fertilized egg, could prevent major damage at the embryonic stage; that such therapy prevents the children of those with a genetic disease from having to undergo somatic cell therapy or from having to make painful reproductive decisions of their own; that germ-line therapy is more cost effective because, unlike somatic therapy that has to be repeated generation after generation, this is done once; that researchers are obligated to identify and develop better treatments to offer to their patients; and finally, that germ-line therapy is a way to prevent serious health problems rather than attempting to repair the damage after it occurs.[17] The major negative arguments are that if there are unforeseen negative side-effects, these will be passed down from generation to generation; that the therapy is not needed, since there are other means to prevent transmitting genetic diseases, such as preimplantation diagnosis or selective abortion following prenatal diagnosis; that germ-line therapy will be expensive and available only to a small number of individuals; that perfecting the methods of germ-line therapy will require much research on human embryos, which many would argue is inappropriate; and finally that, if the technique should prove to be of limited use in curing disease, the focus might shift to the enhancement of one's genetic profile, which would further reduce the number of people who could use the technique.[18]

One other area of philosophical concern here is the status of the inherited human genome. As Maurice de Wachter puts it, "Germ-line gene therapy techniques would violate the rights of subsequent generations to inherit a genetic endowment that has not been intentionally modified."[19] Such a position raises several problems. Is there such a right and what is its basis? Since the human genome continues to be modified through evolution, on what basis is the present form privileged? And how is human dignity harmed if one can intervene to prevent a disease from harming an individual and his or her descendents?

This question also focuses on a particular problem in the debate: What is human nature? An important contribution has been made by W. French Anderson, one of the major scientists involved in human gene transfer, who is also well read in the ethical and philosophical literature. Originally concerned that germ-line intervention could irreversibly change human nature, Anderson has recently argued that the Platonic resolution of human nature into body and soul is correct. Therefore, since the essence of our nature resides in our soul, no bodily alteration can harm human nature. Thus Anderson winds up with a Platonic-Cartesian dualism that sees the body as a *res extensa* with no relation to our human nature or our person. This position is substantively critiqued by James Keenan, S.J., who in a seminal article argues for the subjectivity of the body and who reasons that a separation such as Anderson proposes misunderstands personhood. Keenan also demonstrates the necessity of keeping the body-person at the center of ethical analysis because "recent genetic research substantiates the position that the human body is in its genetic roots profoundly relational and that this position provides substantial guidelines for the genetic manipulation of our progeny."[20] To change the body, therefore, is to change the person. And that is the locus of the next issue, the genetic enhancement of humans.

Will we move beyond therapy to enhance particular human characteristics? A major problem is that no particular single gene has been definitely associated with a particular behavioral characteristic, e.g., intelligence. The enhancement debate is also characterized by an unacknowledged genetic determinism, namely that we can do only as our genes tell us. This assumption that all behaviors—no matter how complex—are caused by a single gene neglects the role of the environment, both physical and social, in developing our characteristics.[21]

Nonetheless, such theoretical arguments will not slow the quest for enhancement, the primary evidence of which is the growing market in sperm, eggs, and embryos from vendors who list their own appearance and their educational and social background, as well as that of their parents and grandparents. Prenatal diagnosis offers another way to select preferred genomes, and as long as parents want both better children and strategies to achieve them, enhancement will be with us. Now the primary method of enhancement is social, through various child-rearing and educational strategies; in the future it may be attempted by selecting desired genotypes. But no matter the means, desire for enhancement brings dangers. In a most interesting discussion of enhancement, Glen Magee has identified five sins of enhancement to avoid: calculativeness, overbearingness, shortsightedness, hasty judgment, and pessimism.[22]

Consideration of some dimensions of human gene transfer brings us back to many of the same issues previously encountered in discussions of genetic testing: human dignity, the extent of human control over nature, understanding of human nature, and our relation to our descendants.

FREEDOM-DETERMINISM AND HUMAN DIGNITY

The categories of freedom-determinism and human dignity show up sharply in the cloning debate. Does our genetic profile determine who we are? Will our acts be determined by our genome? Is our genome our fate?

Four types of cloning must be distinguished lest the debate become even more confused. Gene cloning and cellular cloning are two methods of increasing supplies of DNA or various cells to facilitate experiments; they have nothing to do with whole-organism cloning. A third form of cloning is called blastomere separation or embryo division; it involves artificially twinning an embryo to produce multiple copies. While this form is used in the livestock industry routinely, it has been attempted in humans only in the experiment reported by Hall and Stillman.[23] The fourth type is the one that has occupied center stage since the report of the cloning of Dolly in February 1997.[24] This is somatic cell nuclear transfer, or whole-organism cloning, in which an egg has its nucleus removed and replaced with the nucleus of another cell that produces an identical genetic copy of the donor. When the Dolly story was first made public (the announcement was delayed until three months after her actual birth in order to allow time for the appropriate patents on the technique to be filed), most focused on the application to humans. What had been the stuff of science fiction now appeared to be one more scientific conquest. Yet even in this case most of the debate was misplaced, unless one were a genetic reductionist or determinist. What is of utmost importance in the Dolly experiment is demonstrating that the genetic material in fully differentiated adult cells can be reactivated to generate a whole new being. Such reactivation was previously thought to be impossible (though some now question this because of the impossibility of proving that the cell used for Dolly was in fact an adult cell, though such claims may be ended with the announcement of the cloning of fifty mice, some of which were clones of clones[25]). Second, such a method of reproduction is asexual and occurs without fertilization, hardly the standard way of mammalian reproduction.

Once the news about Dolly was out, the most common scenario imagined the replication of an almost infinite series of desired genotypes on the assumption that they would essentially be the same person—all Michael Jordan clones would be superior basketball players and all James

Watson clones would be superior scientists. There are two major errors in these scenarios. First, the fact that two individuals share the same genetic identity does not mean they are the same person (any more than traditionally conceived identical twins are the same person). Nor does the fact that they share a genetic identity diminish or violate the dignity of either. Second, these scenarios rest on any number of varieties of genetic reductionism that identifies the self with the genome or argues that one's genome alone sets one's life course and all one's choices. Such positions deny any transcendent dimension to the person, any freedom, and simply ignore the role of environment on personal development, either behaviorally or physically. While arguments will continue over the degree of interaction of all these elements, it is clear that the major error of the human cloning debate was genetic reductionism.

Other arguments focused on the violation to human dignity from the process of cloning: not being conceived in the normal fashion, not having two biological parents, not having one's unique genotype.[26] These arguments are not new: they are identical or similar to those raised earlier in discussions of *in vitro* fertilization. And they involve inherently the same problems.[27] Precisely how is human dignity compromised by a conception that is artificially achieved? What is the basis of the asserted right to be conceived "naturally," to be conceived biologically through heterosexual intercourse, or to have two heterosexual parents? Even the position that human life begins at fertilization is impossible to hold, because in cloning there is no fertilization and no sperm. And what is one to think of current research in which "nucleic DNA from several species—rats, sheep, pigs, and rhesus monkeys—[is inserted] into cows' eggs whose own nuclei have been removed, and the eggs activated the nucleic DNA to produce a clone of the donor of the DNA"?[28] If this research is successful, it will solve the problem of the shortage of human eggs for use in assisted reproduction. Thus cloning continues to force the debate over the moral status of the human embryo, and it will heighten the already complex debate over whether or not early human embryos can be created for the exclusive purpose of research or whether or not already-created embryos can be used for that purpose. For if the cloning of humans is to go forward, it must be preceded by some research on human embryos to evaluate both safety and efficacy.

One can distinguish, however, between the means of assisted reproduction and the context of reproduction. If cloning becomes another form of assisted reproduction, it will become another means in a very competitive and lucrative reproductive market. And here the context of reproduction becomes important for moral analysis. First, assisted reproduction is a multimillion dollar per year market, which means that there is keen competition for clients among clinics. Thus there is a strong incentive immediately to implement any new technology that might give one clinic an edge. Andrews reports the statement of a fertility clinician: "We go from mindside to bedside in two weeks. We make things up and try things on patients. We never get their informed consent, because they just want us to make them pregnant."[29] One can hardly expect responsible research on cloning in such a success-driven context. Second, the assumption is that autonomy reigns in this area as in all others in American culture. This of course begs the question whether individual choice is in fact the only morally relevant value in such discussions. Third, and somewhat related, is the assumption that all reproductive choices are private and, therefore, immune from social evaluation. There are social costs to pregnancy that society must bear: higher insurance premiums for plans that subsidize assisted reproduction, increased use of newborn intensive care units resulting from the increase in multiple pregnancies following *in vitro* fertilization and increase and exacerbation of class division between those who can and those who cannot afford the technologies. Fourth are the residual effects and influence of genetic determinism in attempts to custom design children—as the possibilities of selection increase, so too will pressure to select the "best" eggs and sperm from the "best" genetic heritage. Such efforts will create a complex childhood as well as a narrowing of the range of experiences to which a child may be exposed. Growing up has always had its difficulties; growing up with specific expectations grounded in a carefully selected genetic profile may be even more difficult. Finally, it is clear beyond all doubt that we are gaining incredible control over reproduction; the means of reproduction are being instrumentalized. Consequently, we need to keep clearly in mind the larger ends to which these means are being used and the context in which they are being implemented. While I would argue that there is nothing inherently immoral with any form of assisted reproduction, there is a danger that we may lose the sense of a child as a gift and come to look upon children as means to an end, an end that is as carefully designed and programmed as possible. Such social determinism closes a child's future and violates a child's dignity. How we use our powers of reproduction will reveal much about us and our priorities.

The new biology and the new genetics are revealing that medical information (particularly the most intimate details about one's genome) is no longer private. This information has profound consequences for one's employment opportunities, insurance possibilities, and social standing. How this new consciousness will be integrated into traditional American concerns on privacy has yet to be thought through. Similarly, we have yet to evaluate the social risks that information such as this and the development of gene therapies offer. Though one can develop analogies and appeal to a variety of models, one will still not know the impact of information or therapy until they are actually tried. And then, of course, the impact cannot be withdrawn.

The dynamic in American culture has been to do first and question later, if at all. And this tendency may present one of our biggest problems.

A second major issue is the exponentially rapid rate of scientific and technological development. Since the announcement of Dolly, we have also seen the cloning of fifty mice (some of which were clones of clones), the cloning of eight calves, the production of human stem cells from different types of human embryonic tissue, and the claim of using a human-cow embryonic hybrid as another method of developing human stem cells.[30] One can barely keep up with the reports, much less think through the issues. And this pace will continue. A central concern is that many of these developments are produced in private biotech companies that receive no federal funding. This means that there is no necessity of review by an ethics committee or an institutional review board. While some companies have ethical review committees, they are essentially discretionary. Given that research will continue to be controversial as well as complex, we need a way to engage in before-the-fact, responsible discourse over the directions of such research and applications. The Asilomar Conference called by scientists in the wake of the developing recombinant DNA technology in the late 1970s provides a useful model. Perhaps the time for another such conference has come.

NOTES

1. William J. Winslade, "Privacy in Health Care," in *Encyclopedia of Bioethics*, edited by Warren T. Reich, rev. ed. (New York: Macmillan, 1995), vol. 4, 2064–65.

2. For a list and copies of proposed legislation on genetic privacy and confidentiality, see the Web site of The National Human Genome Research Institute, http://nhgri.nih.gov (accessed 21 April 2003); and Philip R. Reilly, "Genetic Privacy Bills Proliferate," *The Gene Letter 1*, http://www.geneletter.org, May 1997.

3. Nancy Kass, "Insurance for the Insurers: The Use of Genetic Tests," *Hastings Center Report* 22 (November–December 1992): 6–11; Thomas H. Murray, "Genetics and the Moral Mission of Health Insurance," *Hastings Center Report* 22 (November–December 1992): 12–17.

4. Thomas H. Murray, "Warning: Screening Workers for Genetic Risk," *Hastings Center Report* 13 (February 1983): 5–8.

5. For overviews and more detailed discussions, see Barbara Katz Rothman, *The Tentative Pregnancy* (New York: Viking, 1986); "Genetic Grammar: 'Health,' 'Illness,' and the Human Genome Project," a special supplement in *Hastings Center Report* 22 (July–August 1992): S11–S20; Edward M. Berger, "Morally Relevant Features of Genetic Maladies and Genetic Testing," in Bernard Gert et al., *Morality and the New Genetics* (Sudbury, Mass.: Jones and Bartlett, 1996); R. Gregg, *Pregnancy in a High-Tech Age: Paradoxes of Choice* (New York: Paragon House, 1993); Larry Thompson, *Correcting the Code: Inventing the Genetic Cure for the Human Body* (New York: Simon and Schuster, 1994); Owynne Basen, Margrit Eicher, and Abby

Lippman, eds., *Misconceptions: The Social Construction of Choice and the New Reproductive and Genetic Technologies* (Quebec City: Voyageur, 1996).

6. For an early framing of the issues, see President's Commission for the Study of Ethical Problems in Medicine and Biomedical and Behavioral Research, *Screening and Counseling for Genetic Conditions* (Washington, D.C.: U.S. Government Printing Office, 1983).

7. For a highly critical view of genetic testing, see Ruth Hubbard and Richard C. Lewontin, "Pitfalls of Genetic Testing," *New England Journal of Medicine* 334 (2 May 1996): 1192–93. But see also Francis S. Collins, "BRCA1—Lots of Mutations, Lots of Dilemmas," *New England Journal of Medicine* 334 (18 Jan 1996): 186–88, which suggests positive strategies for utilizing genetic information on breast cancer; Jerome Groopman, "Decoding Destiny," *The New Yorker* 76 (9 February 1998): 42–48; Albert Rosenfeld, "At Risk for Huntington's Disease: Who Should Know What and When?" *Hastings Center Report* 14 (June 1984): 5–8; A. M. Cordi and J. Brandt, "Psychological Cost and Benefits of Predictive Testing for Huntington's Disease," *American Journal of Medical Genetics* 55 (1995): 618–25; Dorothy C. Wertz et al., "Genetic Testing for Children and Adolescents: Who Decides?" *Journal of the American Medical Association* 272 (1994): 875–81; Sandi Wiggins et al., "The Psychological Consequences of Predictive Testing for Huntington's Disease," *New England Journal of Medicine* 327 (12 November 1992): 1401–05. Abstracts and some articles from the *New England Journal of Medicine* are available at http://www.nejm.org (accessed 22 April 2003). For an excellent British perspective, see *The Troubled Helix: Social and Psychological Implications of the New Human Genetics,* Theresa Marteau and Martin Richards, eds. (New York: Cambridge University, 1996).

8. For example, a 30-year-old woman may gain 2.9 to 5.3 years of life expectancy from prophylactic mastectomy and 0.3 to 1.7 years from prophylactic oophorectomy; see Deborah Schrag et al., "Decision Analysis-Effects of Prophylactic Mastectomy and Oophorectomy on Life Expectancy among Women with BRCA1 and BRCA2 Mutations," *New England Journal of Medicine* 336 (15 May 1997): 1465–71, and the accompanying editorial in the same issue by Bernardine Healy, "BRCA Genes—Bookmaking, Fortunetelling, and Medical Care," 1464.

9. Dorothy Wertz et al., "Genetic Testing for Children and Adolescents," 875–81. Other areas of application are testing children prior to adoption or deciding how to invest one's resources in one's children. Nancy Wexler, president of the Hereditary Disease Foundation and one of the team that discovered the gene for Huntington's disease, reports that a woman asked to have her two children tested for Huntington's because "she had only enough money to send one of them to Harvard" (Mary Murray, "Nancy Wexler," *New York Times Magazine*, 13 February 1994, 31).

10. The reason for this age cut-off is that this is when the risks of having a child with Down syndrome balance the risks of miscarriage from amniocentesis.

11. Cf. Options, http://www.fertility.options.com/, for a sample of the genetic pedigrees that can be ordered from egg and sperm vendors.

12. The single best book on gene therapy is LeRoy Walters and Julie Gage Palmer, *The Ethics of Human Gene Therapy* (New York: Oxford University, 1997). For the reflections and analysis of one of the main proponents and researchers in the field of gene therapy, see W. French Anderson, "Human Gene Therapy: Why Draw A Line?" *Journal of Medicine and Philosophy* (December 1989): 681–93; "Ge-

netics and Human Malleability," *Hastings Center Report* 20 (1990): 21–24; and "Genetic Engineering and Our Humanness," *Human Gene Therapy* 5 (1994): 755–59. For general overviews, see the following: Clifford Grobstein and Michael Flower, "Gene Therapy: Proceed with Caution," *Hastings Center Report* 14 (April 1984): 13–17; Burke K. Zimmerman, "Human Germ-Line Therapy: The Case for its Developments and Use," *Journal of Medicine and Philosophy* 16 (1991): 593–612; Maurice A. M. de Wachter, "Ethical Aspects of Human Germ-Line Therapy," *Bioethics* 7 (1993): 166–77; LeRoy Walters, "Human Gene Therapy: Ethics and Public Policy," *Human Gene Therapy* 2 (Summer 1991): 116–201; David A. Kessler et al., "Regulation of Somatic Cell Therapy and Gene Therapy by the Food and Drug Administration," *New England Journal of Medicine* 329 (14 October 1993): 1169–73; Nelson A. Wivel and LeRoy Walters, "Germ-Line Gene Modification and Disease Prevention: Some Medical and Ethical Perspectives," *Science* 262 (22 October 1993): 533–38; Jeff Lyon and Peter Gorner, *Altered Fates: Gene Therapy and the Retooling of Human Life* (New York: W. W. Norton, 1995).

13. W. French Anderson and T. Friedmann, "Strategies for Gene Therapy," in *The Encyclopedia of Bioethics*, vol. 2, p. 908.

14. Walters and Palmer, *The Ethics of Human Gene Therapy*, 36–44.

15. Jeffrey M. Leiden, "Gene Therapy—Promise, Pitfalls, and Prognosis," *New England Journal of Medicine* 333 (28 September 1995): 871–73.

16. I recall being at a meeting where one of the researchers for the first use of gene therapy was discussing the project. When asked the cost of the research, he replied that he had no idea because the National Institutes of Health (NIH) did not submit a bill. But obviously the NIH has a budget, one that has to be set in relation to other budgets in the health field, to say nothing of other national priorities.

17. Walters and Palmer, *The Ethics of Human Gene Therapy*, 81–82. See also an earlier, but similar, phrasing of the arguments in Eric Juengst, "Germ-Line Therapy: Back to the Basics," *Journal of Medicine and Philosophy* 16 (1991): 587–92. For selected European perspectives on the pros and cons of this debate, see Maurice A. M. de Wachter, "Ethical Aspects of Germ-Line Therapy," *Bioethics* 7 (1993): 166–77.

18. Walters and Palmer, *The Ethics of Human Gene Therapy*, 82–83.

19. de Wachter, "*Ethics of Human Germ-Line Therapy*," 175.

20. James Keenan, S.J., "Genetic Research and the Elusive Body," in *Embodiment, Morality, and Medicine*, Lisa S. Cahill and Margaret A. Farley, eds. (Dordrecht, The Netherlands: Kluwer Academic, 1995): 59–73, at 59.

21. At a meeting attended primarily by scientists to discuss germ-line therapy and possible guidelines for its implementation, James Watson suggested that a serious candidate for a disease to be cured by germ-line therapy was stupidity (Gina Kolata, "Scientists Brace for Changes in Path of Human Evolution," *New York Times*, 21 March 1998, A1 and A7).

22. Glen McGee, *The Perfect Baby: A Pragmatic Approach to Genetics* (Lanham, Md.: Rowman & Littlefield, 1995): 123–33.

23. Jerome L. Hall et al, "Experimental Cloning of Human Polyploid Embryos Using an Artificial Zona Pellucida," a paper presented at the 1993 annual meeting of the American Fertility Society. For an overview of this experiment and its subsequent discussion, see the *Kennedy Institute of Ethics Journal* 4 (September 1994), which devoted the entire issue to this topic. Also Andrea L. Bonnicksen, "Ethical

and Policy Issues in Human Embryo Twinning," *Cambridge Quarterly of Healthcare Ethics* 4 (1995): 268–84.

24. For early discussions see James D. Watson, "Moving Toward Clonal Man: Is This What We Want?" *The Atlantic,* May 1971, 50–53; Martin Ebon, editor, *The Cloning of Man: A Brave New Hope—or Horror?* (New York: New American Library, 1978); Margaret Brumby and Pascal Kasimba, "When Is Cloning Lawful?" *Journal of In Vitro Fertilization and Embryo Transfer* 4 (August 1987): 198–204; Ira H. Carmen, *Cloning and the Constitution: An Inquiry into Governmental Policymaking and Genetic Experimentation* (Madison: University of Wisconsin, 1985). The most comprehensive discussion of cloning can be found in the National Bioethics Advisory Commission, *Cloning Human Beings: Report and Recommendation of the National Bioethics Advisory Commission* (Rockville, Md.: NBAC, 1997); see also the commentary on this report "Cloning Human Beings: Responding to the National Bioethics Advisory Commission's Report," in *Hastings Center Report* 27 (September–October 1997): 6–22; Gina Kolata, *Clone: The Road to Dolly and the Path Ahead* (New York: William Morrow, 1998); Gregory E. Pence, *Who's Afraid of Human Cloning?* (Lanham, Md.: Rowman & Littlefield, 1998). The entire issue of *Cambridge Quarterly of Healthcare Ethics* 7 (1998) is devoted to cloning.

25. Vittorio Sgaramelia and Norton D. Zinder, "Letter to the Editor," *Science* 279 (30 January 1998): 636–66, together with Wilmut's response; see also Gina Kolata, "In Big Advance in Cloning, Biologists Create 50 Mice," *New York Times,* 22 July 1998, A20.

26. See Congregation for the Doctrine of Faith "*Donum vitae,* Instruction on Respect for Human Life in Its Origin and on the Dignity of Procreation" (22 February 1987); this document along with a commentary can be found in Thomas A. Shannon and Lisa S. Cahill, *Religion and Reproduction* (New York: Crossroad, 1988). Such positions are not unique to Roman Catholicism; Paul Ramsey, e.g., argues that any reproductive technology that separates reproduction from heterosexual intercourse is immoral (*Fabricated Man: The Ethics of Genetic Control* [New Haven: Yale University, 1970]).

27. See Edward Vacek, S.J., "Vatican Instruction on Reproductive Technology," *Theological Studies* 49 (1988): 110–31.

28. Lori B. Andrews, "Human Cloning: Assessing the Ethical and Legal Questions," *Chronicle of Higher Education,* 13 February 1998, B4–B5.

29. Ibid., B4.

30. Thomas A. Shannon, "Remaking Ourselves? The Ethics of Stem-Cell Research," *Commonweal* 125 (December 4, 1998): 9–10.

3

Catholic Reflections on the Human Genome

James J. Walter

Religious traditions frequently inform the contexts that shape how believers morally decide complex bioethical topics, and this certainly has been the case for the Roman Catholic community. There is a substantial number of documents from this religious tradition that have been produced on the scientific and medical interventions into the human genome, and the teachings found in these documents have illuminated the way Catholic believers have morally thought about and judged manipulations of the human genetic code.

Before discussing how the Roman Catholic moral tradition has approached the recent discoveries in genetics and the Human Genome Project, however, it would be helpful to name and describe the areas within genetic science that are or will be in need of moral evaluation. Medical scientists could conceivably develop four different types of human gene transfer from the results produced in the Human Genome Project.[1] In other words, within the next twenty or thirty years medical science will have the capability to alter our genetic code in four ways. The first two types are therapeutic in nature because their intent is either to correct some genetic defect that causes disease or to prevent future disease. The other two types are not therapies at all, and many question whether they are part of medicine's goals as well. Rather, they are concerned with improving either various genetic traits of the patient him/herself (somatic cells) or with permanently enhancing or engineering the genetic endowment of the patient's children (germ-line cells).

The first kind of human gene transfer is somatic cell therapy in which a genetic defect in a body cell of a patient could be corrected by using various enzymes (restriction enzymes and ligase) to splice out the defect and to splice in a healthy gene. Medical scientists have already used a variation of this technique to help children who suffer from severe combined immune deficiency (ADA) by modifying bone marrow cells,[2] and a similar procedure was used in August 1999 for children who have Crygler Najjar syndrome, a genetic disease that causes fatal brain damage.[3] Estimates are that between two thousand to five thousand different genetic diseases are controlled by one gene,[4] and these diseases afflict approximately two percent of all live births.[5] Second, there is germ-line gene transfer therapy in which either a genetic defect in the reproductive cells—egg or sperm cells—of a patient would be repaired or a genetic defect in a fertilized ovum would be corrected *in vitro* before it is transferred to its mother's womb.[6] In either case, the patient's future children would be free of the defect by permanently altering their genetic code.

Next there are the two kinds of nontherapeutic, or enhancement, human gene transfer. The first kind is enhancement somatic engineering. In this type, a particular gene could be inserted to improve a normal trait—for example, the insertion of a new gene or an improved one to enhance memory. Second, there is germ-line genetic engineering in which existing genes would be altered or new ones inserted into either germ cells or a fertilized ovum such that these genes would then be permanently passed on to improve or to enhance traits of the patient's offspring. In this last form of human gene transfer, parents could design their children according to their own desires.

In light of these four types of human gene transfer, I will begin with a very general conclusion about the Roman Catholic perspective on genetics and genetic interventions. With the exception of human cloning,[7] *in principle* there is nothing in the Catholic tradition that theologically or morally prohibits interventions into the human genetic code, though in fact there may be circumstances in which a specific intervention might be immoral.[8] To prove this conclusion I will focus my attention primarily on documents from the magisterium, or teaching authority, of the Catholic Church, i.e., documents from recent popes, bishops, and the Second Vatican Council. I will augment these teachings on occasion with various positions taken by Catholic theologians and bioethicists.

I will proceed by analyzing three sets of issues that are at stake for the Catholic tradition on the topic of human gene transfer. These three sets are: (1) anthropological issues, (2) theological issues, and (3) moral issues. My analysis of each must necessarily be brief. Though I realize there is an important dispute about what constitutes a *therapeutic* end and what constitutes an *enhancement* of the human genome,[9] I will limit most of my remarks to the area of the therapeutic, i.e., interventions to cure or to prevent a disease. In the conclusion I will offer a brief summary of some definite positions taken by the Roman Catholic tradition on alterations of the human genome.

ANTHROPOLOGICAL ISSUES

There are several background beliefs about the human that function as starting points for a moral discussion of gene therapy, but I will mention only two. First, the Roman Catholic tradition consistently argues not only that the nature of the human person is both body and spirit but also that there is a oneness among these distinguishable but inseparable aspects. As the *Pastoral Constitution on the Church in the Modern World* (*Gaudium et Spes*) states the matter: "Though made of body and soul, man is one."[10] The current pope, John Paul II, has reiterated this belief in several of his recent statements on genetics. For example, in his 1982 address to the Pontifical Academy of Sciences, the pontiff claimed that, "The human body is not independent of the spirit, just as the spirit is not independent of the body, because of the deep unity and mutual connection that exist between one and the other."[11] Thus, any genetic intervention into the human subject must recognize and respect this unity; any view that separates the two is dualistic and leads to a denigration of one or other of the two aspects of the person.

Second, the Catholic tradition argues that there are various kinds of goods whose pursuit of and acquisition by persons will define their well-being and flourishing. Two of these goods are particularly important here: life and health. In their working report on genetic intervention, the British bishops argued that, "To be fulfilled in our existence as human beings requires some degree of bodily well-being. Health is a good which is a dimension of the basic good of life."[12] Thus, if health is a basic good that all pursue, even though there are definite limits to this pursuit,[13] the nature of this good itself becomes the grounds for the prima facie obligation on the part of both patients and physicians to seek remedies for genetic diseases.[14] The role of medicine, then, is to serve health, and the technological means by which medicine realizes this good are ultimately subject to the objective standards of morality, which themselves are based on the nature of the human person in all its dimensions.[15]

THEOLOGICAL ISSUES

There are two distinctively theological issues that serve as interpretive frameworks for the morality of gene therapy. The position one takes theologically on each of these issues will inform and shape how one reasons morally about genetic interventions.

The first issue concerns a question about whether or not we are "playing God" by intervening in the human genome in order to cure or prevent genetic diseases.[16] This question obviously has definite anthropological implications, because at its core it is asking about the responsibility that we humans should or should not have over material reality, including the materiality (genes) of our own bodies. If the divine has not decided to share with us the dominion over our bodies but has reserved such dominion to itself, then it would seem that any act to change what God has given us in our bodies would be an improper exercise of human freedom and thus an act of "playing God." On the other hand, if one believes that God has indeed granted this responsibility to humans, then it would seem that we have at least a prima facie moral obligation to alter our genetic makeup for therapeutic ends. Such acts in this latter view, then, would not be improper acts of "playing God"; rather, they represent the rightful taking up of our responsibility for the goods of life and health.

For the most part, the current pope,[17] and much of the Catholic tradition,[18] has argued rather strongly for the view that we humans, within certain moral limits, have been granted by the divine the responsibility over material nature, including our own genetic heritage. Consequently, as long as researchers respect the nature of the human person, a moral criterion that I will develop below, at least therapeutic genetic interventions are theologically permitted in the Catholic community.

The second theological issue concerns the question of whether or not a special sacred status should be conferred upon the human genome or DNA either because of its intimate connection to human reproduction and development or because of its participation in the image of God (*imago dei*) that resides in us.[19] If yes, then any intervention into our genetic code would constitute an improper act on the part of medical scientists. If no, then in principle our genes are like all other aspects of the created material world and thus possess no special sacred status. The Catholic tradition has understood the status of the human genome in terms of the latter view. For example, in their document on genetic intervention the British bishops ask if the genome is morally untouchable by virtue of its special role in human development. Their answer,

> We would argue not, in view of the fact that the genome is simply one highly influential part of our bodies: the part which directs the formation of other parts, both in ourselves and in our offspring. We believe that, like other parts of the body, the genome may *in principle* be altered, to cure some defect of the body.[20]

Edmund Pellegrino, M.D., and Joseph Cassidy, O.P., have also argued for a similar position. They claim that human genetic material is a cause of great wonder, but in itself it does not deserve any special status such that interventions into our genome would per se constitute immoral acts.[21] Consequently, from a strictly theological perspective, the Catholic tradition would not prohibit interventions into the human genome for purposes of curing or preventing a genetic disorder.

MORAL ISSUES

Recent documents from the magisterium,[22] especially those from the current pope, reveal a remarkable positive evaluation of genetic interventions. Many of these texts demonstrate an awareness of the difference between somatic cell and germ-line cell interventions. The distinction between therapy and enhancement is acknowledged as well. In principle, none of these in itself is judged morally wrong, but each must be judged according to moral standards. Some of these standards are established moral principles; others serve as the foundation for the moral principles. In what follows, I will list and briefly analyze four of these moral standards in relation to the various types of genetic intervention.

Do Good, Avoid Evil: The Fundamental Moral Imperative

Following Thomas Aquinas's discussion of the natural law in the thirteenth century,[23] this moral standard in the Catholic tradition has been considered the foundation for all moral principles. In the present discussion, the particular goods that we are to pursue are the goods of life and of health. The nature of these goods ground the prima facie obligation to pursue them on behalf of ourselves and on behalf of others. However, we are only strictly obliged to avoid harm; we do not have a strict obligation to accomplish all good.[24] This understanding of our obligations clearly indicates that a good end does not justify a morally bad means and that a strict risk-benefit calculus is not the sole perspective from which to judge the moral appropriateness of genetic interventions.[25]

Genetic Interventions Must Respect the Dignity of the Human Person

This is clearly the most fundamental moral principle that applies to our discussion of genetic intervention, and it takes various forms in the documents under consideration.[26] In its most general terms, science and technology require *for their own intrinsic meaning* an unconditional respect for this principle.[27] Respect must be present from the very moment of conception,[28] and it requires that we not reduce life to a mere object.[29] Scientific interventions into the human genome respect the integrity of the person when they focus on benefits for the patient. Thus, genetic experimentation on human subjects, including embryos, can be justified morally as long as there is informed consent (by the patient or by a proxy) and the experiments avoid harm and are directed to the well-being of the person.[30] Furthermore, experiments that are not strictly directed toward therapy but are aimed at improving the human biological condition (enhancement) can be justified, at least in part, on the grounds that the experiments respect the human person by safeguarding the identity of the person as one in body and soul (*corpore et anima unus*).[31] However, genetic experiments that are directed toward sex selection or other predetermined qualities[32] and those directed toward the creation of different groups of people[33] are forbidden morally because they violate the dignity of the person.

Genetic Interventions Must Promote the Well-Being of the Patient

I have already alluded to this standard above, but it does have the status of a distinct moral principle in the Catholic tradition. John Paul II has used it in part to justify morally the use of therapeutic genetic interventions to cure disease.[34] Likewise, the Science and Human Values Committee of the National Conference of Catholic Bishops has used this principle in permitting genetic testing for a cure or effective therapy of genetic diseases.[35]

Proportion between the Risks and Benefits

This is an important moral principle that applies to this topic, though most of the documents studied reject this as the sole principle that would apply to genetic interventions. The risks and benefits must be calculated in terms of their potential impact upon a patient's well-being and not in terms of their impact on existing others or future humanity.[36] In the end, if the benefits to the patient reasonably outweigh the risks, then this proportion can in part justify genetic interventions.

CONCLUSION

There are a substantial number of documents from the magisterium of the Catholic Church that have been produced on the topic of scientific and medical interventions into the human genome. With the exception of human cloning, for the most part, these teachings have been quite positive in their evaluation of these potential technologies. By way of conclusion, permit me to summarize the results of my analysis in relation to genetics in general and to gene therapy in particular.

First, it seems clear that the Roman Catholic tradition would not only morally permit but would strongly encourage the development of somatic cell therapies as long as these interventions do not violate any of the anthropological, theological, or moral issues discussed previously. In principle, these therapies raise no new moral problems that have not already been dealt with in other types of medical interventions to cure or prevent disease.[37] The goods of life and health ground a prima facie moral obligation in the Catholic tradition to pursue research in this area of therapy.

Second, there seems to be an emerging, but not absolute, consensus that germ-line therapy, if that were ever to become a possibility, would not be considered in principle unacceptable.[38] There are several qualifications that need to be made on this claim, though. It is important to note that this is an *in principle* argument; *de facto* or in practice, germ-line therapy is currently considered unacceptable for several reasons. For example, such therapy would be developed only after experimenting on embryos and exposing them to great harm.[39] In addition, even if one clearly distinguishes gonadal cell germ-line therapy from embryonic cell germ-line therapy, there are still problems. As the Catholic Health Association in the United States has noted, gonadal cell therapy would have to be justified on the grounds of possible beneficial results *for future humanity*, since this type of intervention does not alter the genetic makeup of the "patient."[40] This form of justification seems to violate, at least in a prima facie sense, the moral principle espoused by the Catholic tradition that any intervention should be for the benefit of the patient himself or herself and not for the benefit of some future humanity.

Finally, there is the possibility of genetic enhancement to improve the human. Though there has not been much written on this specific type of genetic manipulation from the perspective of the magisterium, or official teaching authority of the Catholic church, nonetheless at least John Paul II has not ruled it out of hand by declaring it intrinsically immoral. Rather, he seems to be open to such developments as long as they do not violate the moral principles I have already outlined.

NOTES

1. W. French Anderson, "Genetics and Human Malleability," *Hastings Center Report* 20 (1990): 21–24.

2. LeRoy Walters, "Ethical Issues in Human Gene Therapy," *The Journal of Clinical Ethics*, 2 (1991): 270.

3. Pamela Schaeffer, "Special Report: Body and Sold," *National Catholic Reporter* 36 (October 22, 1999): 22. It should also be noted that a similar technique is proposed by the scientists who cloned five piglets on March 5, 2000, in Blacksburg, Va., for xenotransplantation of pig organs to humans. These scientists intend to "knock out" a specific gene responsible for adding a sugar group to pig cells. This sugar group is foreign to the human immune system, so the unaltered organs would be rejected in the human body. Additionally, the scientists would introduce

through gene transfer three new genes into the cells of the cloned pigs to control the causes of organ rejection. See Marjorie Miller, "5 Pigs Cloned: Transplants to Humans Touted," *Los Angeles Times*, March 15, 2000, A1 and A11.

4. Thomas F. Lee, *The Human Genome Project: Cracking the Genetic Code of Life* (New York: Plenum, 1991), 183.

5. Burke K. Zimmerman, "Human Germ-Line Therapy: The Case for its Developments and Use," *The Journal of Medicine and Philosophy* 16 (1991): 595.

6. Brian V. Johnstone, "La tecnologia genética: perspectiva teologico-moral," *Moralia* 2 (1989): 301.

7. There has been a consistent condemnation of human cloning in recent documents from the teaching authority of the Catholic church, and this judgment applies to both embryo splitting and to somatic cell nuclear transplant cloning techniques. For example, see the Congregation for the Doctrine of the Faith, "Instruction on Respect for Human Life in Its Origin and on the Dignity of Procreation" (*Donum vitae*), *Origins* 16 (March 19, 1987): 703; John Paul II, "The Gospel of Life" (*Evangelium Vitae*), *Origins* 24 (April 6, 1995): 711; and idem, "Address to the International Congress on Organ Transplants," http://www.vatican.va/holy_father/john_pa...hf_jp-ii_spe_20000829_transplants_en.html.

8. One Catholic theologian who argued for this conclusion was Karl Rahner, S.J. For two of his most influential essays in this area of genetics, see "The Experiment with Man," in his *Theological Investigations*, vol. IX, trans. Graham Harrison (New York: Herder and Herder, 1972), 205–24; and "The Problem of Genetic Manipulation," ibid., 225–52.

9. For example, see Eric T. Juengst, "Can Enhancement be Distinguished from Prevention in Genetic Medicine?" *The Journal of Medicine and Philosophy* 22 (April 1997): 125–42.

10. *Gaudium et Spes*, in Walter M. Abbott, S.J., editor, *The Documents of Vatican II* (New York: American Press), 212.

11. John Paul II, "Biological Research and Human Dignity," *Origins* 12 (October 22, 1982): 342–43, at 342. Also, see idem, "The Ethics of Genetic Manipulation," *Origins* 13 (November 17, 1983): 385, 387–89, at 387.

12. British Catholic Bishops, *Genetic Intervention on Human Subjects: The Report of a Working Party of The Catholic Bishops' Joint Committee on Bioethical Issues* (London: The Linacre Centre, 1996), 17.

13. In the Roman Catholic theological tradition, to designate any potential medical intervention as "extraordinary," i.e., that there is a disproportion between the benefits and burdens to a patient, would constitute a limit to the obligation that the patient would have to pursue that intervention. For a further discussion of the principle of proportionate versus disproportionate means (ordinary vs. extraordinary means), see the Congregation for the Doctrine of the Faith, "Declaration on Euthanasia," *Origins* 10 (August 14, 1980): 154–57, at 156.

14. The argument that the nature of a good itself grounds the prima facie moral obligation to pursue the good is based on a theory of natural law. For a helpful discussion of natural law theory within the Roman Catholic tradition, see Richard M. Gula, S.S., *Reason Informed by Faith: Foundations of Catholic Morality* (New York: Paulist Press, 1989), chapters 15 and 16.

15. John Paul II, *The Redeemer of Man* (*Redemptor Hominis*) (March 4, 1979), sec. 16. The argument that the objective standards of morality are based on the nature of the

human person originates from the *Pastoral Constitution on the Church in the Modern World* (*Gaudium et Spes*), no. 51. For a very helpful discussion of "the human person integrally and adequately considered," see Louis Janssen's, "Artificial Insemination: Ethical Considerations," *Louvain Studies* 8 (Spring 1980): 3–29.

16. There are many different ways to understand the phrase "playing God." For some it functions as an actual moral judgment on scientific interventions, but for others it serves as a distinctively theological perspective from which to assess these interventions. For a very helpful discussion of these differences, see the Protestant theologian Allen Verhey's, "'Playing God' and Invoking a Perspective," *The Journal of Philosophy and Medicine* 20 (1995): 347–64. Also, see James J. Walter, "'Playing God' or Properly Exercising Human Responsibility? Some Theological Reflections on Human Germ-Line Therapy," *New Theology Review* 10 (November 1997): 39–59; and idem, "Notes on Moral Theology: Theological Issues in Genetics," *Theological Studies* 60 (March 1999): 124–34.

17. For example, see John Paul II, *The Redeemer of Man* (*Redemptor Hominis*), sec. 16.

18. For example, see the Congregation for the Doctrine of the Faith, "Instruction on Respect for Human Life in Its Origin," 699.

19. Though their focus was not on genetic interventions but on the patenting of genes, two Protestant theologians in the Southern Baptist community have recently argued for the special sacred status of genes based on the belief that the image of God (*imago dei*) pervades all aspects of human life, including one's genes. See Richard D. Land and C. Ben Mitchell, "Patenting Life: No," *First Things* 63 (May 1996): 20–23, at 21.

20. British Catholic Bishops, *Genetic Intervention on Human Subjects*, 32. Emphasis in original.

21. Joseph D. Cassidy, O.P., and Edmund Pellegrino, "A Catholic Perspective on Human Gene Therapy," *International Journal of Bioethics* 4 (1993): 11–18, at 12.

22. For example, see Pius XII's 1953 address on genetics, "Moral Aspects of Genetics," in *Medical Ethics: Sources of Catholic Teachings*, 3d ed., edited by Kevin D. O'Rourke, O.P., and Philip Boyle (Washington, D.C.: Georgetown University Press, 1999), 170–71.

23. Thomas Aquinas, *Summa Theologiae*, trans. Fathers of the English Dominican Province (New York: Benzinger Brothers, Inc., 1947), I–II, q. 94, art. 2.

24. Bishops' Committee for Human Values, "Statement on Recombinant DNA Research," in *The Pastoral Letters of the United States Catholic Bishops, vol. IV: 1975–1983*, edited by Hugh J. Nolan (Washington, D.C.: USCC, 1984), 200–04, at 203; and John Paul II, "The Ethics of Genetic Manipulation," 388.

25. Bishops' Committee for Human Values, "Statement on Recombinant DNA Research," 203–04.

26. For example, see *Gaudium et Spes*, no. 51; and the *Catechism of the Catholic Church* (New York: Paulist Press, 1994), 424.

27. Congregation for the Doctrine of the Faith, *Donum vitae*, Introduction, sec. 2.

28. Bishop Friend/Pontifical Sciences Academy, "Frontiers of Genetic Research: Science and Religion," *Origins* 24 (January 19, 1995): 522–28, at 523.

29. See John Paul II, "The Ethics of Genetic Manipulation," 388; and The Catholic Health Association of the United States, *Human Genetics: Ethical Issues in Genetic Testing, Counseling, and Therapy* (St. Louis, Mo.: Catholic Health Association, 1990), 34.

30. *Donum vitae*, sec. 4; and Friend, "Frontiers of Genetic Research," 523.

31. John Paul II, "The Ethics of Genetic Manipulation," 388.

32. Catholic Health Association, *Human Genetics*, 26; and Friend, "Frontiers of Genetic Research," 524.

33. John Paul II, "The Ethics of Genetic Manipulation," 388.

34. Ibid. Also, see idem, "Address to the Members of the Pontifical Academy for Life," February 24, 1998, 1–3. In this address, the pope argues that genetic interventions into somatic cells are permissible because scientists in 1998 have claimed that these procedures are safe. Of course, the fatal genetic interventions performed on Jesse Gelsinger at the University of Pennsylvania had not yet occurred when the pope delivered this address. The pope further asserts in this address that in fact these procedures do not seem safe on germ-line cells or on early embryos, and thus they are not permissible. Also, idem, "Address to the Pontifical Academy of Sciences: The Human Person Must Be the Beginning, Subject and Goal of All Scientific Research," (1994), http://www.its.caltech.edu/~newman/sci-cp/sci94111.html, 3.

35. NCCB/Science and Human Values Committee, "Critical Decisions: Genetic Testing and Its Implications," *Origins* 25 (May 2, 1996): 769, 771–72, at 771.

36. Friend, "Frontiers of Genetic Research," 524.

37. For example, see the Catholic Health Association, *Human Genetics*, 19; British Catholic Bishops, *Genetic Intervention on Human Subjects*, 28; and James F. Keenan, S.J., "What Is Morally New in Genetic Manipulation?" *Human Gene Therapy* 1 (1990): 289–98, at 292.

38. For example, see British Catholic Bishops, *Genetic Intervention on Human Subjects*, 34,

39. Ibid., 42–43. For additional reasons, see the Catholic Health Association, *Human Genetics*, 20–21.

40. Catholic Health Association, *Human Genetics*, 21–22.

4

Reflections on the Moral Status of the Preembryo

Thomas A. Shannon and Allan B. Wolter, O.F.M.

In this paper we wish to review contemporary biological data about the early human embryo in relation to philosophical and theological claims made of it. We are seeking to discover more precisely what degree of moral weight it can reasonably bear. While other ethical conclusions might well be drawn from the results of such a reflective investigation, we limit ourselves to a few moral considerations based on our current knowledge of how human life originates. As Catholics, we too believe that "from the moment of conception, the life of every human being is to be respected in an absolute way because man is the only creature on earth that God 'wished for himself' for himself and the spiritual soul of each man is 'immediately created' by God."[1] But we are also vitally concerned as to when one might reasonably believe such absolute value could be present in a developing organism. We would also like to defuse some of the polar opposition fanned by the rhetoric of both prolife and prochoice advocates that creates a legislative dilemma for morally and religiously responsible politicians. We even hope that a rational analysis of available scientific data might lead to some broad consensus among concerned citizens that the term "human life" is not necessarily a univocal conception.

All life is a many-splendored creation on the part of God; this is especially true of human life at any stage of its development. But we suggest that appropriate protection of the human organism changes with its developmental stages. We wish to present a theory that recognizes the right of every potential mother to a meaningful life and a healthy personality development[2] but that condemns irresponsible destruction of fetal life.

One of the hallmarks of the Catholic tradition, with certain conspicuous exceptions, has been to be in dialogue with the philosophy and science of its day and to use such insights in articulating the vision of Catholicism. Such efforts have been done better and worse. Many have taken time to evaluate the correctness or usefulness of a particular articulation. But in almost all cases, because of new discoveries in science, changes in scientific theory, or the use of new philosophical frameworks, the insights and articulation of the faith of one generation have differed from those of another. Sometimes such differences have led to severe conflict. One remembers the Copernican revolution, the case of Galileo in the seventeenth century, and the tensions introduced by the rediscovery of Aristotelian science in the thirteenth century. Nor can historians of medieval theology forget that certain philosophical views of Aquinas himself were regarded as theologically dangerous by two successive archbishops of Canterbury and condemned by the bishop of Paris in 1277 on the advice of the prestigious university theological faculty, a condemnation that was lifted insofar as it applied to St. Thomas only two years after the saint's canonization in the 14th century.

Anyone who has studied the history of ideas, scientific, philosophical, or theological, knows that there is a usefulness in reviewing the theoretical conceptions of the past, since they have a habit of recurring cyclically in a new and useful scientific garb.[3] The same is true of the theoretic conceptions used by theologians in articulating their faith. We argue that the most recent scientific discoveries fit in more admirably with the epigenetic conception of how a human being originates that was held for centuries by the great theologians and doctors of the Church than does the more recent and now more commonly accepted—though happily not defined—moment of fertilization as coincident with the time of animation. The widespread acceptance of the theory of immediate animation is of post-Tridentine origin,[4] having entered into the tradition only in the early seventeenth century, and in 1869 the distinction between the formed and unformed fetus was no longer canonically recognized. This assumption about immediate animation still plays a large part in contemporary ecclesiastical documents, as well as references to the scientific literature purporting to buttress arguments supporting the theory, as we will discuss later.

We would also like to remind our readers, however, that some forty years ago two learned priests from the University of Louvain,[5] where this theory of immediate animation was originally introduced, repudiated its scientific standing and went to some lengths to explain historically how this mistaken interpretation of empirical data was initially accepted. We claim that the most recent scientific evidence concerning fertilization and the development of the very early human embryo does even more to reinforce their view that any theory of immediate animation seems to have become as untenable today as it was commonly held to be for centuries by Catholic thinkers. We think that since scientific observations, now recognized as erroneous, played such a historical role in the development of the position favoring such a theory, new and respected scientific evidence should be utilized by Catholic theologians when they discuss the process of fertilization and conception to determine its moral implications.

We hope our analysis will be welcomed because of our acceptance and use of the methodology of the tradition and because we take seriously the role of science in helping articulate the context of moral problems, as do current ecclesiastical documents. While our conclusions may differ from those of these documents, we think such differences are to be cherished because they help the community understand its beliefs and values at a much deeper level and allow some of the forgotten riches of our Catholic tradition to be expressed to a new audience.

This rearticulation needs careful examination, however, for the fact that something is new does not *ipso facto* make it good or correct. Thus a careful and prayerful process of discernment should also be an important part of the way we rearticulate our tradition, for the community must genuinely receive the reconceptualization of the tradition before it is authentic. This essay is an attempt at such a process of discernment by setting out an account of the process of individuation in the early human embryo in light of modern biology and reflecting on it in the light of some important theological and philosophical insights that seem to have perennial vitality.

The medieval and post-Renaissance theologians articulated their theory of the person, the body, and ensoulment in light of the biology and philosophy of their day. On the basis of this they appropriately drew moral conclusions. We know now that the biology used at any one time, if not out of date, may well need updating. But the philosophy and history of science also make it clear that there is a significant difference as to how our scientific knowledge of the wonder of God's creation grows. We believe that such a moment of review is necessary today if we are to give a reasonable defense of the respect Catholics have traditionally had for human life. For we know that in the male seed there is no homunculus, but it was not until the 1700s that mammalian sperm was discovered, and not until the 1800s that the mammalian egg was found and its role revealed. Modern diagnostic technologies such as ultrasound and fetoscopy have given us a whole new perspective on the development of the human embryo. Thus, while we can correctly say that the biological data of a past era are inadequate in light of the discoveries of modern science, we cannot dispose as easily of the basic philosophical or theological way our scholastic predecessors interpreted those data. And we certainly cannot fault their use of the most advanced scientific knowledge available to them as a necessary condition for articulating any rational philosophico-theological conception of the person, the body, and ensoulment. It is in that spirit that we present this brief review of what embryology has to tell us today.

CONTEMPORARY PERSPECTIVES ON THE HUMAN EMBRYO

The Preembryo[6]

In mammalian reproduction an egg and sperm unite to produce a new and almost always genetically unique individual. The process, how this occurs, is undergoing tremendous reconceptualization and remodeling in the light of new studies and new diagnostic technologies that allow access to this entity.

A critical discovery of the past two decades is that of capacitation, "the process by which sperm become capable of fertilizing eggs."[7] Human sperm need to be in the female reproductive tract for about seven hours before they are ready to fertilize the egg. This process removes, or deactivates, "a so-called decapacitating factor that binds to sperm as they pass through the male reproductive tract."[8] This permits the acrosome reaction to occur, which is the means by which lytic enzymes in the sperm "are released so that they can facilitate the passage of the sperm through the egg coverings."[9] Then the sperm are able to penetrate the egg so fertilization can begin.

Fertilization usually occurs in the end of the Fallopian tube nearest the ovary. Sperm usually take about ten hours to reach the egg, which if not "fertilized within 24 hours after ovulation, it dies."[10] Fertilization, however, is not just a simple penetration of the surface of the egg. Rather, it is a complex biochemical process in which a sperm gradually penetrates various layers of the egg. Only after this single sperm has fully penetrated the egg and the haploid female nucleus—one having only one chromosome pair—has developed, do the cytoplasm of the egg and the nuclear contents of the sperm finally merge to give the new entity its diploid set of chromosomes. This process is called syngamy. It takes about twenty-four hours to complete and the resulting entity is called the zygote. Thus the process of fertilization (and it is important to note that it is a process) generally takes between twelve to twenty-four hours to complete,[11] with another twenty-four-hour period required for the two haploid nuclei to fuse.

Fertilization accomplishes four major events: giving the entity the complete set of forty-six chromosomes; determination of chromosome sex; the establishment of genetic variability; and the initiation of cleavage, the cell division of the entity.

Now begins a very complex set of cell divisions as the fertilized egg begins its journey down the Fallopian tube to the uterus. About thirty hours after fertilization, there is a two-cell division; after about forty to fifty hours, there is a division into four cells; and after about sixty hours the eight-stage cell division is reached. "When the embryo approaches the entrance to the uterus, it is in the 12-16 cell stage, the morula. This occurs on the fourth day."[12] Although the cells become compacted here, there is yet no predetermination of any one cell to become a specific entity or part of an entity. On around the sixth or seventh day the organism, now called the blastocyst, reaches the uterine wall and begins the process of its implantation there so that it can continue to develop. Here we have a differentiation into two types of cells: the trophectoderm, which becomes the outer wall of the blastocyst, and the inner cell mass, which becomes the precursor of the embryo proper. This process of implantation is completed by the end of the second week, at which time there is "primitive utero-placental circulation."[13]

Critical to note is that from the blastocyst state to the completion of implantation the preembryo is capable of dividing into multiple entities.[14] In a few documented cases these entities have, after division, recombined into one entity again. Nor must this particular zygote become a human; it can become a hydatidiform mole, a product of an abnormal fertilization which is formed of placental tissue.

Note also that the zygote does not possess sufficient genetic information within its chromosomes to develop into an embryo that will be the precursor of an individual member of the human species. At this stage the zygote is neither self-contained nor self-sufficient for such further development, as was earlier believed. To become a human embryo, further essential and supplementary genetic information to what can be found in the zygote itself is required, namely

> the genetic material from maternal mitochondria, and the maternal or paternal genetic messages in the form of messenger RNA or proteins. In terms of molecular biology, it is incorrect to say that the zygote has all the informing molecules for embryo development; rather, at most, the zygote possesses the molecules that have the potential to acquire informing capacity.[15]
>
> That potential informing capacity is given in time through interaction with other molecules. . . . This new molecule with its informing capacity was not coded in the genome. Thus, the determination to be or to have particular characteristics is given in time through the information resulting from the interaction between the molecules.[16]

The development of the zygote depends at each moment on several factors: the progressive actualization of its own genetically coded information, the actualization of pieces of information that originate *de novo* during the embryonic process, and exogenous information independent of the control of the zygote.

The Embryo

The next major stage of development is that of the embryo. This is the beginning of the third week of pregnancy and "coincides with the week that follows the first missed menstrual period."[17] This phase begins with the full implantation of the preembryo into the uterine wall and the development of a variety of connective tissues between it and the uterine wall. Eventually the placenta develops and is the medium through which maternal-embryonic exchanges occur.

Two major events now occur. The first is the completion of gastrulation, "profound but well-ordered rearrangements of the cells in the embryo."[18] This process results in the development of various layers that ultimately give rise to the tissues and organs of the entity and is completed by the third week. At this time all expressions of the genes are switched off except those that determine what a particular cell will be. There are now three layers present that are responsible for the development of much of the organism:

> The embryonic ectoderm gives rise to the epidermis; the nervous system; the sensory epithelitim of the eye, ear, and nose; and the enamel of the teeth. The embryonic endoderm forms the linings of the digestive and respiratory tracts. The embryonic mesoderm becomes muscle, connective tissue, bone and blood vessels.[19]

The second major event, the process of embryogenesis, or organogenesis, now begins and is completed by the end of the eighth week. This process results in the development of all major internal and external structures and organs.

By the end of the third week the primitive cardiovascular system has begun to form with the development of blood vessels, blood cells, and a primitive heart. Since the "circulation of blood starts by the end of the third week as the tubular heart begins to beat,"[20] the cardiovascular system reaches a functional state first.

The nervous system progresses from a neural tube to the essential subdivisions of the brain into forebrain, midbrain, and hindbrain.[21] During this time also the upper and lower limb buds begin to appear. The digestive tract begins to form, as do all the external structures such as the head and the eyes and ears. Hands and feet make their appearance, as do, by the end of the eighth week, distinct fingers and toes.

The development of the nervous system is critical because this is the basis for the "generation and coordination of most of the functional activities of the body."[22] The rudimentary brain and spinal cord are present around the third week but are as yet "unspecialized or undifferentiated for neural function."[23] Neuron development begins around the fifth week, and around the sixth week the "first synapses . . . can be recognized."[24] Carlson observes that at about the seventh week "the embryo is capable of making weak twitches in the neck in response to striking the lips or nose with a fine bristle."[25] Grobstein notes, "the earliest continuous neuronal circuitry for reflex conduction and behavior could be initiated as early as six weeks."[26] Such a pattern, Carlson says, "signifies that the first functional reflex arcs have been laid down."[27]

In a rather thorough review of the literature Michael Flower describes various embryonic movements and the neural basis necessary for their possibility.[28] Flower notes that the earliest reported elicited reflex response from an embryo occurred at 7.5 weeks. This was a movement away from a stroking stimulus to the mouth. Such movements were typical during this period of the eighth week of development.[29] In the middle of the ninth week the patterns make a transition to whole body responses, and during the twelfth week local reflexes dominate. These data indicate a critical level of integration of the nervous system.

This review of embryonic development up to the eighth week shows a dramatic process of development from the initiation of fertilization to the formation of an integrated organism around midgestation. The rest of the paper will concentrate on examining what moral implications these data might have. The intent is not to draw a moral *ought* from a biological *is*, but to reconsider the compatibility of moral and philosophical claims with what we know of developmental embryology.

MORAL CONSIDERATIONS

Conception

A critical finding of modern biology is that conception biologically speaking is a process beginning with the penetration of the outer layer of the egg by a sperm and concluding with the formation of the diploid set of chromosomes. This is a process that takes at least a day. This raises a question as to how one ought to understand the term *moment of conception* frequently used in church documents.

One could understand "moment" metaphorically as referring to the process as a whole, or if it is meant to convey an instant of time, then it would seem to refer to either the end of the process of biological conception when the zygote has become an embryo, or to some prior stage of development that has been reached in which this human life form (fertilized egg, zygote, or preembryo) has acquired a distinct set of properties. However, it seems that the theologians who framed these carefully crafted documents wished to convey the idea that at the moment of conception (whatever stage of development of human life obtains) everything is present that is required essentially for this human organism to be a person in the philosophical or theological, if not psychological, sense of the term: a rational or immortal soul has been created and infused into the organic body. At the same time, while they wished to set forth guidelines, they declared it was still a theoretically open question and hence they did not want to specify, or define, the moment when such passive conception (as it was called by Catholic theologians for many centuries) took place. Prayerful reflection on what embryology and our Catholic tradition tell us may not yield any direct positive knowledge of when passive conception takes place, but it does seem to throw considerable light on when it has not occurred.

Biologically understood, conception occurs only after a lengthy process has been completed and is more closely identified with implantation than fertilization.[30] The pastoral letter *Human Life in Our Day* speaks of conception "initiating . . . a process whose purpose is the realization of human personality."[31] Such a phrase is biologically correct if applied to implantation and seems to be a reasonable moral description of the typical outcome of conception.

Singleness

Clearly and without any doubt, once biological conception is completed we have a living entity and one that has the genotype of the human species. As Grobstein nicely phrases it, "conception (fertilization) is the beginning of a new generation in the genetic sense."[32] This zygote is capable of further divisions and is clearly the precursor of all that follows. But can we say with *Donum vitae*, quoting the "Declaration on Procured Abortion," "From the time that the ovum is fertilized, a new life is begun which is neither that of the father nor of the mother; it is rather the life of a new human being with his own growth?"[33]

How are we to understand this phraseology in the light of the biology of development? For, while it is correct to say that the life that is present in the newly fertilized egg is distinct from the father and mother and is in fact usually genetically unique, it is not the case that this particular zygote is fully formed and it is not a single human individual, an "ontological individual," as Ford suggests.[34] Because of the possibility of twinning, recombination, and the potency of any cell up to gastrulation to become a complete entity, this particular zygote cannot necessarily be said to be the beginning of a specific, genetically unique individual human being. While the zygote is the beginning of genetically distinct life, it is neither an ontological individual nor necessarily the immediate precursor of one.

Second, the zygote gives rise to further divisions "resulting in an aggregate of cells, each of which remains equivalent to a zygote in the sense that it can become all or any part of an embryo and its extraembryonic structure."[35] Such cells at this stage are totipotent:

> Within the fertilized ovum lies the capability to form an entire organism. In many vertebrates the individual cells resulting from the first few divisions af-

> ter fertilization retain this capability. In the jargon of embryology, such cells are described as totipotent. As development continues, the cells gradually lose the ability to form all the types of cells that are found in the adult body. It is as if they were funneled into progressively narrower channels. The reduction of the developmental options permitted to a cell is called restriction. Very little is known about the mechanisms that bring about restriction, and the sequence and time course of restriction vary considerably from one species to another.[36]

Such a process of restriction is completed when the cells have become "committed to a single developmental fate. . . . Thus determination represents the final step in the process of restriction."[37] Such determination begins during gastrulation, three weeks into embryonic development.

Genetic uniqueness and singleness coincide on one level only after the process of implantation has been completed and on another after the restriction process is completed. Thus, if we take implantation as the marker of both conception and human singleness, this does not occur until about a week after the initiation of fertilization. If we use determination and restriction, because of their signaling of the loss of totipotency of the cells, as the markers of human singleness, then individuality does not occur until about three weeks after fertilization. Of critical importance is Ford's observation: "The teleological system of the blastocyst should not be identified with the ontological unity of the human individual that will develop from it."[38]

There is, then, a partial answer to the very interesting question[39] *Donum vitae* asks: "How could a human individual not be a human person?"[40] A Catholic philosopher might well object or reply that this is certainly a very muddied question, for "traditionally speaking" individuality has been considered a necessary, though not sufficient, condition for human personhood. The rational soul has never been considered the formal reason why something human is individual. Obviously, "human individual" can have several meanings. If it refers to a fertilized ovum, this is indeed something both human (qua product) and numerically single. Yet, until the process of individuation is completed, the ovum is not an individual, since a determinate and irreversible individuality is a necessary, if not sufficient, condition for it to be a human person.

Something human and individual is not a human person until he or she is a human individual, that is, not until after the process of individuation is completed. Neither the zygote nor the blastocyst is an ontological individual, even though it is genetically unique and distinct from the parents. The potential for twinning remains until the beginning of gastrulation, although it is rare for it to occur this late. Additionally, a zygote that divides can reunite and one individual will emerge. Furthermore, each cell can form a total individual. A human individual, to use the language of the document, cannot be a human person until after individuality is established.

Also, as Grobstein noted, genetic uniqueness does not necessarily imply singleness.[41] That is, when fertilization is complete and the diploid state is reached, the organism has its full complement of genetic information. At this point it is genetically unique. But because of the potentiality for twinning, this uniqueness may be shared by more than one organism. Thus, even though unique, the organism is not necessarily single. Singleness or individuality occurs after the genetically unique organism has implanted and its development is restricted to forming one unified organism.

An individual is not an individual, and therefore not a person, until the process of restriction is complete and determination of particular cells has occurred. Then, and only then, is it clear that another individual cannot come from the cells of this embryo. Then, and only then, is it clear that this particular individual embryo will be only this single embryo.

One can reasonably conclude, then, that if there is no single human entity, there is no person. For the one is the presupposition of the other. Thus, when *Donum vitae* approvingly refers to the findings of modern science and argues "that in the zygote . . . resulting from fertilization the biological identity of a new human individual is already constituted,"[42] does not this statement of the congregation fail to make a critical distinction between genetic uniqueness and singleness? In using "individual" rather than "person" in this meticulously worded statement, the congregation may have sought to sidestep the controversial question of when personhood begins. But if "individual" be taken in its philosophical or technical meaning, scientific data available today hardly justify the claim that a particular zygote is necessarily both genetically unique and an individual.

This is particularly important in assessing the theological intent of the congregation, particularly since it argues that the "conclusions of science regarding the human embryo provide a valuable indication for discerning by the use of reason a personal presence at the moment of this first appearance of a human life."[43] As the statement stands, three concepts appear to be conflated here: genetic uniqueness, singleness, and personal presence. The argument for the first presence of human and personal life in the zygote relies heavily on scientific claims about the fertilized egg. However, such claims of singleness and personhood cannot be made, the former scientifically and the latter philosophically. We assume that the congregation would want to adjust its findings in the light of these distinctions.

Ensoulment[44]

In this section and elsewhere, we will be discussing the principle of immaterial individuality or immaterial selfhood. In the Catholic tradition, and clearly in many of the sources we cite, the usual term for this is "soul." Our practice will be to use the term "soul" when speaking within a clear traditional context. But when we develop our own presentation, we will use the term "immaterial individuality" or "immaterial selfhood," because the term "soul" has many connotations and images connected with it, and in so far as possible we wish to avoid problematic usages and confusing images.

Issues

Although far from being a defined doctrine, there is support in Roman Catholic moral theology for the position that ensoulment is coincident with fertilization or, at least, as early as possible after conception. This position apparently dates from the early seventeenth century writings of Thomas Fienus, professor on the faculty of medicine at Louvain.[45] This opinion gradually caught on and became the dominant opinion. This position was complemented by teachings that held that the embryo "possesses the essential parts of a human body, though very minute in size."[46] This teaching on immediate animation eventually worked its way into the mainstream of Catholic moral theology. If doctors of medicine were Catholics, explains Dorlodot,

> they were told that the theologians of their time held that the soul is created by God immediately after fecundation. The theologians in turn based themselves on the opinion of the doctors, as these did on that of the theologian. In other words, *caecus caeco ducatum praestat*. Finally, the moral theologians, who completely forgot the principles, which, according to the great doctors of Catholic morality, render abortion always illicit, invoked the danger of favouring abortive or sterilizing practices.[47]

Additionally, the removal from canon law in 1896 of the distinction between the formed and unformed fetus suggests that there is not a time when the body is unformed.[48] The *Ethical and Religious Directives for Catholic Health Facilities* provide another reason when they include in the definition of an abortion the "interval between conception and implantation."[49] Also, we have the 1981 testimony of Cardinal Cooke and Archbishop Roach in support of the Hatch amendment: "We do claim that each human individual comes into existence at conception, and that all subsequent stages of growth and development in which such abilities are acquired are just that—stages of growth and development in the life cycle of an individual already in existence."[50] Finally, in *Donum vitae* we read: "nevertheless, the conclusions of science regarding the human embryo provide a valuable indication for discerning by the use of reason a personal presence at the moment of this first appearance of a human life."[51]

If this statement is to be accepted as it stands, we suggest that the conclusions of science should be interpreted differently, particularly if we reflect on what we know from science in the light of a centuries-long tradition among Catholic philosophers and theologians. For like them we are struck by both the wonder and sacredness of human life even from its obscure beginnings, as well as to when we could begin to suspect a personal presence might be there. Nor can we forget that for some seventeen centuries the Church indeed condemned abortion, but not on the ground that it might by even the most remote possibility be in all cases a question of murder. Certainly some of the greatest minds and doctors of the Church refused to believe, as many today seem to do, that ensoulment is coincident with fertilization or that we must trace the genesis of each human person back to that moment. Obviously, the Sacred Congregation for the Doctrine of Faith had no intention of definitively settling this question, for it stated pointedly, "This declaration expressly leaves aside the question of the moment when the spiritual soul is infused. There is not a unanimous tradition on this point and authors are as yet in disagreement."[52] It did not believe, however, that such theoretical openness should lead to any rash or precipitous practical action, for it goes on to say, "From a moral point of view this is certain: even if a doubt existed concerning whether the fruit of conception is already a human person, it is objectively a grave sin to dare to risk murder."[53]

Several very critical questions arise here, particularly since abortion was traditionally considered a sin against marriage but not homicide. One of them, concerning the moral possibility of acting on probable knowledge, has already been masterfully treated by Carol Tauer.[54] Others concern practical and philosophical issues relating to the development of the preembryo and embryo. It is to these issues that we now turn.

The dominant position of the moral tradition on ensoulment was the acceptance of a time during the pregnancy when the fetus was not informed by the rational soul. Two distinctions were used in discussing this. The first distinction is between active and passive conception and is exemplified in *De testis* of Benedict XIV, in which the pope comments on the doctrine of the Immaculate Conception.

> Conception can have a twofold meaning, for it is either active, in which the holy parents of the Blessed Virgin, joining each other in a marital role, have accomplished those things which have to do most of all with the formation, organization, and disposition of the body itself for receiving a rational soul to be infused by God; or it is passive, when the rational soul is coupled with the body. This infusion and union of the soul with a duly organized body is commonly called passive conception, namely, that which occurs at that very instant when the rational soul is united with a body consisting of all its members and its organs.[55]

Thus the pope would seem to understand active conception, in our terminology, as the physical union of egg and sperm that will become the embryo, while passive conception would be the moment the rational soul is infused into a suitably organized body, one that results from (begins with) organogenesis.

The second distinction is between mediate and immediate animation by such a soul. The theory of mediate animation is succinctly stated as follows:

> Animation by the intellectual soul is impossible so long as the parts of the brain which are the seat of the imagination and the vis cogitativa (and we might add the memory) are not suitably organized. But it still is more evident that there cannot he animation by the intellectual soul when the brain is not even outlined, or again, when even the embryo really does not as yet exist. Now that is precisely the case with the ovum, and the morula, and of that which results from its development, so long as there has not appeared, on a particular part of the germ, that which by its ulterior development will become a fetus.[56]

Immediate animation occurs coincidentally with the fusion of egg and sperm, known as the moment of conception. This is the position utilized in the teachings referred to at the beginning of this section. This distinction is also thoroughly discussed by Donceel, as previously noted."[57]

Medieval theologians were particularly interested in clarifying the technical meaning of "conception" in their justification of the celebration of the popular feast of the Blessed Virgin Mary's conception. Henry of Ghent, following common scholastic reasoning, distinguished between the "conception of the seed when fetal life begins" and the conception of the human soul some "35 or 42 days later [when], depending on the sex, a rational soul is created."[58] Such a position echoes St. Anselm's perceptive judgment, "No human intellect accepts the view that an infant has a rational soul from the moment of conception."[59]

Had this saint known of the empirical data on wastage, he would have considered such a claim not only irrational but blasphemous.[60] For only about 45 percent of eggs that are fertilized actually come to term. The other 55 percent miscarry for a variety of reasons. Some are related to the biochemistry of the uterus, others are a function of low levels of necessary hormones, while yet other reasons have to do with structural anomalies within the preembryo or embryo itself.[61] Such vast embryonic loss intuitively argues against the creation of a principle of immaterial individuality at conception. What meaning is there in the creation of such a principle when there is such a high probability that this entity will not develop to the embryo stage, much less come to term?

Also, given the fact that twinning and recombination is a possibility, what is one to say about the presence of immaterial individuality during that process? If this principle is initiated at fertilization and then a twin is formed, how does one explain the relation of the original principle to the zygote that splits off? And should recombination occur, how does one explain coherently the fate of such a principle of immaterial individuality?

Should one freeze the preembryo, all organic processes stop for the duration. What is the status of immaterial individuation then? It is genuinely unclear what to think of that in terms of the standard theory of immediate ensoulment. Then there is the issue of whether a soul, in the classic sense of the form of the body, is needed for the fertilized egg to develop into its possible subsequent forms.

Commentary

The question of the moral significance of the morula and of embryonic wastage has been noted previously in the moral literature. In 1976, for example, Bernard Haering brought together much of the scientific literature and examined its moral significance. His conclusion concurs with one suggestion in our analysis and opens the door to other issues: "the argument that the morula cannot yet be a person or an individual with all the rights of the members of the human species seems to me to be convincing as long as we follow our traditional concept of personhood."[62] This conclusion opens up several areas for consideration.

First, we concur with Haering and particularly with the analysis of Ford that, given the biological evidence, there is no reasonable way in which the fertilized egg can be considered a physical individual minimally until after implantation. Maximally, one could argue that full individuality is not achieved until the restriction process is completed and cells have lost their totipotency. Thus the range of time for the achievement of physical individuality is between one and three weeks. One simply cannot speak, therefore, of an individual being present from the moment of fertilization.

Second, given the standard definition of personhood used in Catholic moral theory—an individual substance of a rational nature—questions are raised about the rational nature. When might one consider such a rational nature to be present? Ford suggests the formation of the primitive streak, which coincides with the time of the formation of the neural tube, as an appropriate criterion.[63] Another criterion would be around eight weeks, when the first elicited responses have been recorded. These are the result of a simple three-neuron circuit. Thus, toward the end of the embryonic period some neural activity is present. A third answer would be the formation of a relatively integrated nervous system, which occurs around the twentieth week of fetal development. Of critical importance here is the connection of neural pathways through the thalamus to the neocortex. This allows stimuli to be received, as well as activities to be initiated.

One can speak of a rational nature in a philosophically significant sense only when the biological structures necessary to perform rational actions are present, as opposed to only reflex activities. The biological data suggest that the minimum time for the presence of a rational nature would be around the twentieth week, when neural integration of the entire organism has been established. The presence of such a structure does not argue that the fetus is positing rational actions, only that the biological presupposition for such actions is present.

Third, the preembryonic form as a system is not totally passive, the recipient only of actions from the outside as it were. It has its own activities arising from the released potencies of the novel combination of its constituent materials. Such potencies are released when these elements form a system, e.g., the embryo. This development of new systems gives rise to new activities and possibilities and serves as the foundation or presupposition for other stages of development. Philosophically speaking, we have every reason to believe that the dynamic properties of the organic matter—the elements of the fully formed zygote—owe their existence to their organizational form or the system. Important to note is that "where there are only material powers—that is, the ability to form material systems—there is only a material nature or substance."[64] Thus the material system or form of the developing body can explain its own activity. We conclude that there is no cogent reason, either from a philosophical or still less from a theological viewpoint, why we should assert, for instance, that the human soul is either necessary or directly responsible for the architectonic chemical behavior of nucleo-proteins in the human body.

Among the scholastic theologians and doctors of the Church, perhaps St. Bonaventure has given the most helpful model for what we have in mind. For in his interesting Aristotelian interpretation of how St. Augustine's theory of seminal reasons might be explained according to the science of his own day, he argued that if the potencies be understood as active rather than passive, then the Aristotelian formula that the new substantial form *is educed from the potency of matter* made sense. For "the philosopher of nature says that matter first receives the elementary form and by its means it comes to the form of the mineral compound only by means of the latter to the organic form, for he looks to that potency of matter according to which it is progressively actualized by the operation of nature."[65]

If we interpret this in more contemporary terms, it means simply that the new substantial form is nothing more than that of the organic system itself, and that its new and unique dynamic properties stem from the complementary interaction of elements that make up the system. All that is needed is some external agent to bring the elements of that system together, for, as Bonaventure puts it, "in matter itself there is something cocreated with it from which the agent acting in matter educes the form. Not that this something from which the form is educed is such that it becomes some part of the form to be produced, but it is rather that which can be and will become the form, even as a rosebud becomes a rose."[66]

These remarks suggest that the principle of immaterial individuality is indeed the ultimate actualization of all the potencies contained within the forms or systems that constitute the organic life of the human being. Thus, finally, we can say that while it is necessary to recognize the distinctions between higher and lower vital functions in the human being, nonetheless there may be "an area where the biochemical theory is the more plausible explanation, and another area where the animistic position seems to be the only tenable view."[67]

The question of when such a principle comes into being is dependent on which level of the system of the human being one is examining and what activities are performed here. The strong implication of these suggestions is that immaterial individuality comes into existence late in the development of the physical individual.

CONCLUSIONS

Biological Data

Physical Individuality

Two biological data mandate a revision of our understanding of the beginning of individuality: (1) the possibility of twinning, which lasts up to implantation, which occurs about a week after fertilization begins, and (2) the completion of the restriction process, which prevents individual cells from forming another individual, about three weeks into the pregnancy. While one can speak of genetic uniqueness, in that the fertilized egg has its own genetic code distinct from any other entity (except an identical twin, triplet, etc.), we simply cannot speak of an individual until in fact that individual is present, and the earliest that can be is about two or three weeks after fertilization begins.

Neural Development

Three markers are significant in neural development: (1) gastrulation, the development of the various layers in the preembryo that give rise to the whole organism; (2) organogenesis, the presence of all major systems of the body, occurring around the eighth week; and (3) the development of the thalamus, which permits the full integration of the nervous system, around the twentieth week.

Critical here is the necessity of a functioning and probably integrated nervous system for the possibility of rational activity. For if there is no nervous system functioning, it is not clear that the rational part of the definition of a person can be fulfilled, even though the individual part might be. The functioning nervous system is a necessary condition for the possibility of a new stage of development to emerge and is also a sign that the organism is prepared for this. Thus any of the three markers noted immediately above could serve as an indicator of the capacity for rationality though not necessarily its actuality.

Developmental Autonomy

Given the philosophical discussion on nature and substance, it is reasonable to argue that the developing body as an organized system is a new substance or nature and has the capacity to elicit the potencies within its own reality. That is, a fully formed zygote is a new nature because it has its own actuality and potentiality. It is in itself a sufficient explanation of its own development and activities. The same is true on each new level of development as the zygote becomes an embryo and, finally, a fetus. On a genetic level, the clearest marker of the presence of self-directing activity that would manifest such a new nature would appear in the zygote after it developed the capacity to manufacture its own messenger DNA and thus be developmentally, though not physically, independent of the mother.

Moral Implications

Physical Individuality

We find it impossible to speak of a true individual, an ontological individual, as present from fertilization. There is a time period of about three weeks during which it is biologically unrealistic to speak of a physical individual. This means that the reality of a person, however one might define that term, is not present at least until individualization has occurred. Individuality is an absolute, or necessary, condition for personhood.

We conclude that there is no individual and therefore no person present until either restriction or gastrulation is completed, about three weeks after fertilization. To abort at this time would end life and terminate genetic uniqueness, to be sure. But in a moral sense one is certainly not murdering, because there is no individual to be the personal referent of such an action.

Since the zygote is living, has the human genetic code, and indeed possesses genetic uniqueness, this entity is valuable, and its value does not depend on the presence or absence of any or a particular quality or characteristic such as intelligence or capacity for relationships.[68] Thus the zygote and the blastomeres derived from it, because they are living, possess ontic value and are in themselves valuable. Thus the general argument made here is not a so-called quality of life argument.

Nonetheless, until the completion of restriction or gastrulation, the zygote and its sequelae are in a rather fluid process and are not physical individuals and therefore cannot be persons. The preembryo at this state, we conclude, cannot claim absolute protection based on claims to personhood grounded in ontological individuality. Yet, since the preembryo is living and possesses genetic uniqueness, some claims to protection are possible. But these may not be absolute and, if not, could yield to other moral claims.

Immaterial Individuality

If one assumes, as we think correct to do, that the potencies actualized in the formation of the new nature of the fertilized egg have the inherent capacity to ground its growth and development, then there is no need to posit a principle of individual immateriality, understood as the Aristotelian nous or as the entelechy of the body, in preembryonic development.

Since the evidence for such a principle comes from the internal evidence of those who experience it, it is difficult at best to ground any speculation as to when it comes into existence. We would make this argument. On the one hand, the developing preembryo as a new nature has within it the potential for future development. On the other hand, if the will as a rational potency is what genuinely distinguishes the person from a nature, then one needs to look to biological presuppositions that enable such a potency to exist. We would argue that the earliest time is around the eighth week of gestation, because then the nervous system is fully integrated.

SUMMARY

We have reviewed some of the salient biological data about the initial stages of the development of human life, with a view to evaluating the philosophical and theological claims made of them. Reflecting on these from a historico-theological perspective, we have tried to discover whether there exists some rational justification for the absolute value that is attributed to the zygote or preembryonic state based on claims to personhood, or whether our earlier long-standing Catholic tradition of mediate animation by a rational soul does not provide a more satisfactory philosophical and theological account. For if we consider judiciously what the great scholastic doctors had to say about the "moment of conception," we seem to have good reason to reintroduce, in interpreting the data of present-day science, the theological distinction between active and passive conception made by Pope Benedict XIV in discussing Mary's immaculate conception.

We thus affirm that any abortion is a premoral evil. That is, it is the ending of life. Consequently, we do not want to be understood as proposing or supporting an "abortion on demand" position or assuming that early abortions are amoral. Abortion is a serious issue, because life is involved and one needs always to respect life. We have made one major argument, however, in this essay. Given the findings of modern biology, there is no evidence for the presence of a separate ontological individual until the completion of either restriction or gastrulation, which occurs around three weeks after fertilization. Therefore, there is no reasonable basis for arguing that the preembryo is morally equivalent to a person or is a person as a basis for prohibiting abortion. That is, there is no biological support for the position that the fertilized egg is from the beginning of the process of fertilization a distinct individual needing no outside agency to develop into a person. Neither is there good philosophical evidence that the principle of immaterial individuality need be present from the beginning to explain the physical development of the preembryo.

This position obviously does not support the argument that abortion is to be prohibited because a person is present from the beginning of fertilization. The earliest such an argument could reasonably be made is after the completion of gastrulation. We recognize that this argument will dismay many and comfort others. Our intention in proposing the argument of this essay is to gain a greater coherence between moral theology and modern embryology.

In this sense we are complementing the work of the Roman congregations and bringing it up to date. We also wish to test the strength of our argument, already subjected to review by several colleagues, in review by a wider and more diverse audience. Additionally, our intention is to develop a position that is reasonable and can be reasonably defended in the public sector.[69] Finally, we think our position on the preembryo and embryo can stand rigorous scrutiny, and we propose it as a factor in developing a feasible stand or national policy on abortion.

One is reminded here of Henry de Dorlodot's evaluation of immediate animation made over fifty years ago in his seminal work *Darwinism and Catholic Thought*: "We are not exaggerating in the least when we regard the fact that this theory [of immediate animation] should still find defenders long after the experimental bases on which it was thought to be founded have been shown definitely to be false, as one of the most shameful things in the history of thought."[70]

NOTES

1. Cf. *Donum vitae*, quoting *Gaudium et spes*, in Thomas A. Shannon and Lisa Sowle Cahill, *Religion and Artificial Reproduction* (New York: Crossroad), 147.

2. We are concerned here especially with victims of rape, incest, or sexual abuse.

3. Philosophers of science have stressed the important difference between the linear growth of scientific data and theoretic conceptions used to interpret them, for important theories have a life of their own that ensures their perenniality. Or, as Santayana put it, those who forget history are condemned to repeat its mistakes.

4. For theologians at the Council of Trent, in contrasting the virginal conception of Christ with the ordinary course of human nature, asserted that normally no human embryo could be informed by a human soul except after a certain period of time "cum servato naturae ordine nullum corpus, nisi intra prasecriptum temporis spatium, hominis anima informari queat" (*Catechism of the Council of Trent*, Part 1, art. 3, n. 7) cited in E. C. Messenger, *Theology and Evolution* (Westminster, Md.: Newman, 1949), 236.

5. We refer to Dr. Messenger and Canon Henry de Dorlodot.

6. This is the term being used to describe this entity from the zygote state to the beginning of the formation of the primitive streak during the third week (see Keith L. Moore, *Essentials of Human Embryology* [Philadelphia: Decker, 1988], 16). The primitive streak gives rise to other structures that continue the physical development of the embryo. The purpose of using this term, as well as other terms such as zygote, embryo, and fetus, is to integrate scientific descriptions into the moral discussion. These terms, as used in this essay, beg no moral questions but help us clearly identify the entity we are discussing. Cf. Clifford Grobstein, *Science and the Unborn: Choosing Human Futures* (New York: Basic Books, 1988), 62. But see *Donum vitae,* which also uses these terms but attributes "to them an identical relevance in order to designate the result (whether visible or not) of human generation from the first moment of its existence until birth" (introduction, 1, n.). The text of *Donum vitae* can be found in Shannon and Cahill, *Religion and Artificial Reproduction,* 140ff. All references will be to this text.

7. Steven B. Oppenheimer and George Lefever Jr., *Introduction to Embryonic Development*, 2d ed. (Boston: Allyn and Bacon, 1984), 87.

8. Ibid., 87.

9. Bruce M. Carlson, *Patten's Foundations of Embryology*, 5th ed. (New York: McGraw-Hill, 1988), 134.

10. Oppenheimer and Lefever, *Embryonic Development,* 175.

11. Ibid., 176.

12. Ibid., 175.

13. Moore, *Essentials*, 14.

14. Carlson, *Patten's Foundations*, 35.

15. Carlos A. Bedate and Robert C. Cegalo, "The Zygote: To Be or Not To Be a Person," *Journal of Medicine and Philosophy* 14 (1989): 642–43.

16. Bedate and Cegalo, "The Zygote," 644.

17. Moore, *Essentials*, 16.

18. Carlson, *Patten's Foundations*, 186.

19. Moore, *Essentials*, 18.

20. Ibid., 24.

21. Carlson, *Patten's Foundations*, 296.

22. Ibid., 456.

23. Grobstein, *Science*, 47.

24. Ibid., 48.

25. Carlson, *Patten's Foundations*, 457.

26. Grobstein, *Science*, 48.

27. Carlson, *Patten's Foundations*, 458.

28. Michael J. Flower, "Neuromaturation of the Human Fetus," *Journal of Medicine and Philosophy* 10 (1985): 237–51.

29. Ibid., 238–39.

30. Norman M. Ford, *When Did I Begin? Conception of the Human Individual in History, Philosophy and Science* (Cambridge: Cambridge University, 1988), 176–77. This outstanding and comprehensive analysis of the biological data came to our attention after we had completed much of our own research for this article. We wish to acknowledge how much we have learned from it and to commend it for its exceptionally thorough review of the biological data and philosophical analysis. We also wish to acknowledge the earlier contribution of James J. Diamond, M.D., to this topic, "Abortion, Animation, and Biological Hominization," *Theological Studies* 36 (1975): 305–24.

31. *Human Life in Our Day*, par. 84.

32. Grobstein, *Science*, 25.

33. *Donum vitae* I, 2, in Shannon and Cahill, *Religion and Artificial Reproduction*, 148.

34. An ontological individual is defined as a "single concrete entity that exists as a distinct being and is not an aggregation of smaller things nor merely a part of a greater whole; hence its unity is said to be intrinsic," Ford, *When Did I Begin?* 212.

35. Grobstein, "Early Development," 235.

36. Carlson, *Patten's Foundations*, 23.

37. Ibid., 26.

38. Ford, *When Did I Begin?* 158. Italics ours.

39. Although any conclusions should not be laid at his door, Richard McCormick, S.J., started Shannon thinking about this problem and was suggestive in phrasing the question.

40. *Donum vitae* I, 2, in Shannon and Cahill, *Religion and Artificial Reproduction*, 149.

41. Grobstein, *Science*, 25.

42. *Donum vitae*, I, 2, in Shannon and Cahill, *Religion and Artificial Reproduction*, 149.

43. Ibid.

44. There is much literature on this, but two interesting articles that are extremely useful for their summaries are Joseph Donceel, S.J., "A Liberal Catholic's View," in *Abortion in a Changing World*, edited by Robert E. Hall (New York: Columbia University Press, 1970); and Carol Tauer, "The Tradition of Probabilism and the Moral Status of the Early Embryo," *Theological Studies* 45 (1984): 3–33. Both articles can be found in *Abortion and U.S. Catholicism: The American Debate*, edited by Patricia B. Jung and Thomas A. Shannon (New York: Crossroad, 1988).

45. Henry de Dorlodot, "A Vindication of the Mediate Animation Theory," in *Theology and Evolution*, edited by E. C. Messenger (Westminster, Md.: Newman, 1959), 271.

46. Ibid., 273.

47. Ibid.

48. Cf. John Connery, S.J., *Abortion: The Development of the Roman Catholic Perspective* (Chicago: Loyola University Press, 1977), 212.

49. The United States Catholic Conference, Washington, D.C., 4.

50. Archbishop John Roach and Cardinal Terence Cooke, "Testimony in Support of the Hatch Amendment, *Origins* 11 (1981): 357–72; also in Jung and Shannon, *Abortion*, 15.

51. *Donum vitae*, I, 1, in Shannon and Cahill, *Religion and Artificial Reproduction*, 149.

52. *Declaration on Abortion* (Washington, D.C.: U.S. Catholic Conference, 1975), 13.

53. Ibid., 6.

54. See n. 44, above. While many have been unhappy with Carol Tauer's article and have dismissed it, Shannon has not yet seen a substantive refutation of her argument that the "application of the probabilist methods would permit some early abortion." Jung and Shannon, *Abortion*, 79.

55. "Conceptio dupliciter accipi potest: vel enim est activa, in qua Sancti B. Virginis parentes opere maritali invicem convenientes, praestiterunt ea quae maxime spectabant ad ipsius corporis formationem, organizationem et dispositionem ad recipiendam animam rationalem Deo infundendam; vel est passiva, cum rationalis anima cum corpore copulatur. Ipsa animae infusio et unio cum corpore debite organizato vulgo nominatur Conceptio passiva, quae scilicet fit illo ipso instanti quo rationalis anima corpori omnibus membris ac suis organis constanti unitur." Benedict XIV, *De festis*, lib. II, chap. 15, n. 1 in *Opera Omnia* 9, edited by J. Silvester (Prato: Aldina, 1843), 303a.

56. Dorlodot, "A Vindication," 266. It was here that Messenger and Dorlodot recalled that the only theological attempt to define the role of the rational soul as the substantial form of the body was made by the council of Vienne (DS 481) and that the fathers and theologians of that council did not subscribe to the immediate animation theory. Dorlodot uses the definition of the council as the major premise of his argument vindicating the mediate animation theory. See Messenger, *Theology and Evolution*, 259.

57. Donceel, "A Liberal Catholic," 48ff.

58. *Quodlibet* 1, 5, g. 13; cited in *John Duns Scotus: Four Questions on Mary*, translated and introduction by Allan B. Wolter, O.F.M. (Santa Barbara, Calif.: Old Mission, 1988), 6. It is interesting to note that Henry breaks with the tradition and ascribes a longer period of gestation before animation to the male rather than the female as was customary since Aristotle.

59. Anselm of Canterbury, *De conceptu virginali et de originali peccato*, chap. 7 in *Anselmi Cantuariensis archiepiscopi opera omnia* 2, edited by F. S. Schmitt (Stuttgart: Bad Cannstatt, 1968), 148 (Anselm of Canterbury 3d ed., translated by Jasper Hopkins and Herbert Richardson [Toronto: Edwin Mellen, 1976], 1–152). It is important to keep in mind that the Archbishop of Canterbury was arguing as to when it is possible to contract original sin, something that all theologians in his day agreed required only the existence of a human soul, not any consciousness or voluntary activity on the part of an infant. As he puts it, "Either from the very moment of his conception an infant has a rational soul (without which he cannot have a rational will) or else at the moment of his conception he has no original sin. But no human intellect accepts the view that an infant has a human soul from the moment of his conception. For from this view it would follow that whenever—even at the very moment of conception—the human seed perished before attaining a human form, the [alleged] human soul in this seed would be condemned, since it would not be

reconciled through Christ—a consequence which is utterly absurd." Today we may have different conceptions as to the nature of original sin and how it is contracted, but we have even less reason than Anselm to believe that there is the remotest possibility of a human will present in what he calls "human seed" at the moment the zygote is formed, or that there is any less rather than a substantially greater amount of "human seed that perishes before attaining a human form."

60. Those who see no insuperable difficulty for the theory of immediate animation in the fact that twins can come from a single fertilized egg should find considerable difficulty in the problem of wastage. To ascribe such bungling of the conception process to an all-wise creator would seem almost sacrilegious. One would have to assume that God in His foreknowledge would create souls only for those He foreknew would eventually be born, an argument a prochoice advocate might well apply to aborted fetuses. On the other hand, Catholics, on the basis of rational argument, can hardly hope to argue for anything more than a suitable level of protection warranted by the development stage of the preembryo and its sequelae.

61. C. Grobstein, M. Flower, and J. Mendeloff, "External Human Fertilization: An Evaluation of Policy," *Science* 222 (October 14, 1983): 127–33.

62. Bernard Haering, "New Dimensions of Responsible Parenthood," *Theological Studies* 37 (1976): 127. This article is also a good review of the scientific literature of that period and contains references to other articles that discuss our theme.

63. Ford, *When Did I Begin?* 171ff.

64. Allan B. Wolter, O.F.M., "Chemical Substance," in *Philosophy of Science* (Jamaica, N.Y.: St. John's University, 1960), 108. This citation is an excerpt from a seminal article originally titled "The Problem of Substance." Its primary aim was to present a cosmological account of how mechanical and natural systems differ, why various forms of living substances arise from nonliving matter, and how traditional scholastic philosophical insights and theories such as both the pluriform and uniform hylomorphic conceptions might be helpful as partial insights to a more complex philosophical theory. The psychological role of the rational soul was discussed peripherally to show how medieval scholastics fitted it into their theories of mediate animation.

65. See J. F. Wipple and Allan B. Wolter, O.F.M., *Medieval Philosophy: From St. Augustine to Nicholas of Cusa* (New York: Free Press and Collier Macmillan, 1969), 325.

66. Ibid., 320.

67. Wolter, "Chemical Substance," 126.

68. For a further discussion of the concept, see James J. Walter, "The Meaning and Validity of Quality of Life Judgments in Contemporary Roman Catholic Medical Ethics," *Louvain Studies* 13 (1988): 195–208. Another discussion can be found in Thomas A. Shannon and James J. Walter, "The PVS Patient and the Forgoing/Withdrawing of Medical Nutrition and Hydration," *Theological Studies* 49 (1988): 623–47.

69. We suggest that something of the violence between extreme prolife or proabortionists might be defused, and the political dilemma of Catholic politicians seeking more rational options might be solved, if one were to recognize that the moral status of, and hence the protection appropriate for, a fetus changes with its developmental stage.

70. Quoted by Messenger, *Theology and Evolution*, 219.

5

Human Nature in a Postgenomic World

Thomas A. Shannon

If anything would generally characterize our current situation, it is the prefix *post* attached to an ever-growing number of nouns to form an adjective describing our world, our civilization, and our relation to them. Among the first *post* generation in more modern times was the post-Galileo generation that experienced the decentering of the earth in its vision of the solar system. Another *post* culture was that of the post-Reformation with both the affirmation of religious freedom and the rise of nation-states, each with its own religious identity. Then came the post-Darwinian culture with its removal of humanity from the apex of the great Platonic chain of being and the striking of a near lethal-blow to hierarchy, both biological and social. Perhaps more significant was the consequent introduction of the concept of change into our notion of reality. For Darwinian thought did provide a devastating, if not fatal, blow to the tree of stability, or stasis. Another contribution to the *post* civilization was that of Freudianism that decentered our concept of the self from both its medieval and Enlightenment position of ahistorical privilege and located it in the midst of a struggle for dominance with the forces of the id. Not only is evolution present in the species but also within the bosom of each human. Currently we have postmodernity with its affirmation of process, dynamism, and the decentering of the text as well as the self, resulting in almost boundless reconstructions of text and self.

Given all these seismic cultural shifts, one would think we might be entitled to a period of integration or at least recuperation from the challenge of making sense of all this. Such is not the case. We are now the postgenomic age that will be the recipient of the fruits of the completion of the Human Genome Project (HGP). While the current focus of the HGP has been its medical implications, the HGP also has implications for our understanding of ourselves, our very human nature, and our relation to others with whom we share our genome, as well as those whose genome differs from ours by perhaps only 3 or 4 percentage points.

The story of the HGP begins of course with the discovery in 1953 of the structure of the DNA molecule by Watson and Crick and continued through the next decades with one discovery after another almost at the proverbial warp speed introduced by the popular TV series *Star Trek*. Such discoveries also gave us the capacity, in the words of the same show, to go where no one had gone before. Now we are on another voyage of self-discovery, a part of which will be very difficult for it will involve leaving a comfortable harbor or at least a known harbor. But another part of the voyage may be even more difficult—the reconstruction of a new vision of human nature in light of our new and ever increasing understanding of the human genome. As the great American philosopher Woody Allen has noted, while the unexamined life may not be worth living, the examined life is no bowl of cherries either!

Before taking some first steps of this journey, I want to make some comments about methodology. With respect to the HGP, much of its success, as well as the success of science in general, is due to the method of reductionism. This method succeeds by breaking components into ever-smaller units and examining them. The whole is explained in terms of the parts and their interaction. This method has been and will continue to be extremely successful and is not to be rejected. A point I would stress is not to confuse the method with a philosophy. That is, to argue that one needs to understand the workings of an organism by understanding its parts—its genetic structure, for example—is not necessarily to argue that an understanding of the genetic structure is a sufficient explanation of the operations of the organism as a whole. One can commit oneself to the use of reductionism as a method without necessarily committing oneself to a philosophy of materialism. This point will recur throughout this paper and I wanted to highlight it here.

A second point is what is referred to in Roman Catholic tradition as *ressourcement*, a method developed by German and French theologians in the 1950s that sought to reappropriate concepts and ideas from the tradition and apply them to or use them to illuminate contemporary discussions.[1] This is not a matter simply of a language change or a method of "We used to say that, but now we say this, but it really doesn't make any difference because both really mean the same." I want to affirm that while our reality is different, particularly given the substantive cultural shifts we have experienced, insights and ideas from the tradition may provide a different angle of vision or bring a critical question to a contemporary discussion. I am not arguing that we can impose the conceptual framework of the past on the present. Rather I am seeking to bring the best of the past with me as I seek to understand what we share with so many people, past and present: our human nature. And part of that nature is surely our past, both genetically and culturally.

Finally, the recently completed HGP has given us a map of the human genome. We now know the location of most individual genes, and the next task is to learn the function of these genes and their interaction with each other and the environment. Only when we begin to understand this dimension of our genetic structure will we be in a better position to achieve a more critical understanding of ourselves. But until then, and I want to emphasize this strongly, we are at the level of knowing the location of the genes and the biological or medical function of only a few of these genes. In spite of all the articles and hype that surround the routine announcement of a gene for this or a gene for that, very little of the actual effects of a particular gene, gene-gene, or gene-environment interaction is actually known. This is particularly the case when the behavior involved is a very complex one such as intelligence, sexual preference, or aggression. Thus at present, we can make only limited comments about human nature based on information from the genetic map we have at our disposal. What we do have, however, are perspectives from current developments in genetics as well as synthetic perspectives such as sociobiology. Even though sociobiology is quite controversial—both with respect to the theories themselves and the perspectives of the critics—information from this field, combined with some information from current genetics, points us in various directions and gives us important information to consider.

To ask the question of the nature of human nature, then, is to enter a whole series of philosophical, scientific, and, for some, theological questions. It is also to enter the complexity of the disciplinary issues within each of these general disciplines and the internal disputes endemic to each. Then there is the problem of any sort of integration of one's knowledge and the validity of the methodological claims on which one rests the validity of such integration.

To choose a context is to choose a viewpoint and to choose a viewpoint is to choose not to see from other viewpoints. This does not mean that other viewpoints are invalid or wrong, but that a multiplicity of viewpoints cannot be simultaneously maintained. This is why the metaphor of triangulation from biological, philosophical, and cultural perspectives is, I think, critical in this essay. One needs to think of human beings and human nature from a variety of viewpoints so that one can eventually gain some perspective and some overlap of perspective. By sighting ourselves from different perspectives, we can gradually gain a deeper understanding of our nature. In particular in this essay, I will be focusing on issues of freedom, altruism, and transcendence because of their centrality in both philosophical and biological discussions.

But to do this is to enter into a variety of controversies: creationism vs. evolution, the sociobiology wars, the mechanisms of evolution debate, philosophical debates, and theological controversies. I think this cannot be avoided. Simultaneously, we must also be aware of the provisional nature of our method and argument. Today's commonly accepted facts are tomorrow's erroneous theories.

In what follows, then, I wish to present several perspectives on developments in contemporary genetics to help learn who we are as humans as well as what implications these perspectives might have for understanding our place in our common cosmos, as well as the implications for religion and ethics.

1. GENERAL PERSPECTIVES ON HUMAN NATURE FROM GENETICS

Evolution

Although perhaps something like 30 to 40 percent of Americans and the school board of the state of Kansas might disagree, the dominant scientifically accepted explanation of the development of life on this planet—from viruses to humans and everything in between—is some form of Darwin's theory of evolution. This theory has been united with elements of Mendelian genetics to form what is referred to as the Modern Synthesis. Part of this agenda is to explain the precise mechanisms of evolution—population genetics, kin selection, adaptationism, punctuated equilibria, sociobiology—but another part of the agenda is to understand the implications of these explanations for understanding ourselves and how we behave, in short, understanding human nature.

A major battle in the 1950s, for example, was the implicit prohibition of hereditarian explanations for human behavior and a focus on cultural or social explanations. The cultural explanation was given official status by the "UNESCO agreement in 1952, which effectively put a ban on biological research in human behavior."[2] Socially, Segerstråle relates this to the influence of immigrant groups in the United States and the Great Depression that made establishing a relation between economic success and biological fitness harder to maintain. Additionally, the anthropologists Franz Boaz, Ruth Benedict, and Margaret Mead made a successful argument for the prominence of culture over biology. Finally, biological or hereditarian explanations of differences were seen as racist, a view made easier by the excesses of the uses of genetics in Nazi Germany as well as at least the rhetoric of the eugenics movement.

However in the 1980s, genetic or behavioral explanations gained ascendancy, a position for which Segerstråle gives several reasons. A major share of credit for this goes to the HGP that focused attention again on the role of genetics. Additionally, the field of anthropology focused on commonalties of human behavior rather than diversity, and this gave more credence to some biological explanations. Language was understood as an adaptive response rather than a purely cultural artifact. And we humans are more frequently described as being in continuity with animals than before, with the emphasis on nonverbal communication and emotions, particularly the emotion of morality.[3]

A second shift is in the perspective on genetics: from nature-nurture to gene and environment to gene-environment (including culture) interaction. The critical issue here is a shift from the role of single genes and their frequency in a population or their random recombination (in which evolution is mainly an additive phenomenon) to a perspective that sees multifaceted feedback loops between and among genes and their environment—a perspective that highlights the complexity of the interaction as well as decreases the role of single genes (except for some diseases).[4]

The shift over the last decades focused on the complexity of the makeup of organisms. Even synthetic approaches such as sociobiology appreciate the complexity of the organism and the critical interaction between its genome and the environment in which it exists. Ironically, as a result of the mapping of the human genome, we seem to be in danger of a return to a genetic essentialism or a focus on the single gene. Some have used the results of this map to emphasize the role of the single gene for determining particular diseases, traits, or behaviors, regardless of their complexity. Thus in addition to constant announcements of discoveries of genes for any number of diseases, we also have the concomitant announcement of single or a small number of genes for complex behaviors such as homosexuality, alcoholism, intelligence, shyness, aggression, and all manner of other behaviors. We seem to be returning to an earlier genetic essentialism, a genetic explanation of behavior that focuses exclusively on the role of the single gene rather than gene-gene interaction or the interaction of the genome as a whole with the larger environment. This often-unacknowledged shift will have profound implications for how we understand ourselves, and we need to keep this perspective in mind as we think about human nature.

Biological Solidarity

One of the most critical discoveries of modern genetics is the commonality of the DNA of all organisms. This biological solidarity is extremely interesting as well as quite threatening. Studies of mammals, primates, other vertebrates, as well as other organisms reveal a striking complementarity of genetic structure. It is clear that humans differ genetically from orangutans and other chimps by perhaps only 1 or 2 percent. The mouse is becoming a major model for the study of human diseases because their genetic profiles overlap considerably.

The question is whether to focus on differences or solidarity. Obviously the differences are critical and 1 or 2 percent of DNA in the right place and in relation to specific environments does make a critical difference as the history of human culture reveals. As Jonathan Marks notes: "The fact that our DNA is 98 percent identical to that of a chimp is not a transcendent statement about our natures, but merely a decontextualized and culturally interpreted datum."[5] Thus by looking at both chimps and humans we can differentiate them quite easily as well as spot several common characteristics. As Marks notes: "The apparent paradox is simply a result of how mundane the apes have become, and how exotic DNA still is."[6]

A critical question emerging from both solidarity and diversity is: do shared genes act differently in humans than in other mammals? This of course is one of the key questions in the sociobiology wars, for E. O. Wilson defines sociobiology as the "systematic study of the biological basis of all social behavior" and suggests a high degree of continuity between mammalian and human behavior. But on the other hand, Wilson exhibits a degree of ambiguity in his argument. For example, he states that genes hold culture on a leash. However Segerstråle notes that Wilson suggests the possibility of aggression's being a recently acquired trait in which a "learned behavior may be 'tracked' genetically. Here, then, we may have the protostatement of his famous pronunciation that 'the genes hold culture on a leash'—this time run in the *opposite* direction, however; that is: culture holding the *genes* on a leash, or the genes tracking culture."[7] And then there is the famous sentence of Dawkins' book *The Selfish Gene*: "We alone on earth, can rebel against the tyranny of the selfish replicators."[8]

Is animal behavior a model or predictor of human behavior? How do we understand the term altruism as applied to animals and humans? Finally we have the question implicitly raised by Dawkins: if we can rebel against our genes, what is the basis for this?

Race and Human Origins

One of the causes of contention among humans has been the phenomenon of racism. The perception of the superiority of a set of physical characteristics, a specific trait, or even the assumption of the possession of a superior genotype has been the source or cause of racism, war, public policy, and much individual and social pain and sorrow. The perception of advantage has been the cause of enormous grief. However, contemporary biology and genetics have taught us something very important: "the careful study of hidden variations, unrelated to climate, has confirmed that homogeneous races do not exist. It is not only true that racial purity does not exist in nature; it is entirely unachievable, and would not be desirable."[9]

On the other hand, it is clear that groups differ from each other, for example, with respect to skin color, eye shape, hair texture, height, etc. Cavalli-Sforza argues that the primary explanation of such characteristics is environmental. He gives four arguments. First, "exposure to a new environment inevitably causes an adaptation to it." Variations in skin color as well as body shape and size are adaptations to temperature and humidity. Second, "there is little climatic variation in the area where a particular population lives, but there are significant variations between the climates of the Earth. Therefore, adaptive reactions to climate must generate groups that are genetically homogeneous in an area that is climatically homogeneous, and groups that are very different in areas with different climates." Third, "adaptations to climate primarily affect surface characteristics." Fourth, "we can see only the body's surface, as affected by climate, which distinguishes one relatively homogeneous population from another."[10] Others note that perhaps .01 percent of our genes are responsible for our external appearance and that we differ "from one another only once in a thousand subunits of the genome."[11]

It is clear that there are many differences between humans and human groups. But these differences do not constitute a race: "a group of individuals that we can recognize as biologically different from others."[12] Such differences would have to be biological and statistically significant. "Because genetic divergence increases in a continuous manner, it is obvious that any definition or threshold would be completely arbitrary."[13] And while such information might logically demonstrate the uselessness of classification—for example, efforts to establish some sort of superiority—Cavalli-Sforza does indicate one justification for genetic classification: to identify groups with a genetic similarity that, because of common ancestry, increases the probability of having similar diseases and, therefore, the possibility of developing drugs responsive to these diseases. Here the motive for classification is therapeutic and justified by the humanitarian need to cure disease.

Another argument against the concept of race and racism is the common origin of all modern humans from a population in Africa. The separation of chimps and humans occurred about five million years ago, and modern humans arose in Africa about 100,000 years ago. The age of this so-called African Eve, or more precisely mitochondrial Eve, was calculated by counting "the number of mutations that differentiate two living individuals, and identify when their last common ancestor lived."[14] Such calculations gave rise to the notion of an African Eve: "the woman whose mitochondria were the last common ancestors of all surviving mitochondria today" and who lived around 190,000 years ago.[15] A similar African Adam was found by developing techniques to trace nucleotide mutations of the Y chromosome, and this African Adam's age was dated at around 144,000 years ago. Thus modern genetics, in addition to modern anthropology, demonstrate what seems to be a significant human reality: "the continents were settled by Africans in the expected order. Modern humans appear first in Africa, then in Asia, and from this big continent they settled its three appendices: Oceania, Europe, and America."[16] Such migrations began 100,000 to 80,000 years ago. And as the populations grew and migration occurred, so did the process of adaptation to new environments and climates that in turn led to the differences we currently observe between and among modern humans. Such differences are environmental adaptations by groups but are not genetic and cannot override the reality of our common origin and cannot provide any justification for any claims to superiority, genetic or otherwise.

Individuality

Populations are essentially homogeneous with some variations being a function of distance from the original ancestor. But even these differences slow down as geographic distance increases. Scientifically then, it is irresponsible to use the term race to denote some sort of biological superiority or the primacy of some genotype or some group. However, such homogeneity is not the case in looking at individuals. The argument for this comes from the various technologies involved in DNA fingerprinting, which identifies the probability of a DNA specimen coming from a particular individual. "The chance of two [unrelated] individuals on average having the same DNA profile is about one in a million billion" according to one researcher in the forensic application of this technology.[17] And as Cavalli-Sforza notes: "Regardless of the type of genetic markers used (selected from a very wide range), the variation between two random individuals within any one population is 85 percent as large as that between two individuals randomly selected from the world's population."[18] Additionally, through migration and increasing intermarriage, we have a greater mixing of genes that will have two effects: first, decreasing any genetic differences between groups and, second, increasing the differences between individuals of the same population.

The medieval champion of individuality John Duns Scotus anticipated something of Cavalli-Sforza's insight into the significance of individuality:

> In the universe as a whole, order is mainly considered according to types or species where their inequalities or differences pertain to order. According to Augustine, however, in the *City of God* (XIX, chapter 13) 'order is an arrangement of like and unlike things whereby each of them is disposed in its proper place.' That is why this Agent who primarily intended the order of the universe (as the principle good, intrinsic to Himself) not only intended this in-

> equality that is one requirement for order (among species) but also desired a parity of individuals (within the same species) which is another accompaniment of order. And individuals are intended in an unqualified sense by this First One insofar as he intended something other than himself not as an end, but as something oriented to that end. Hence to communicate his goodness, as something befitting his beauty, he produces several in each species. And in those beings which are the highest and most important, it is the individual that is primarily intended by God.[19]

Conclusions

This general orientation lays out some critical insights into our considerations of human nature from the perspective of modern genetics. We are a dynamic, evolving species with a common genetic and a common geographical origin. We have a genome that is very adaptive and responsive to a variety of environments. Cavalli-Sforza neatly summarizes this:

> Anthropometric characteristics, including skin color, demonstrate the selective effects of the different climates to which modern humans have been exposed in the course of their migrations over the Earth's surface. They vary especially with latitude. By contrast, genes are considerably more useful as markers of human evolutionary history, especially migrations. They vary more with longitude.[20]

The differences between populations are skin deep and essentially irrelevant socially or politically. However, within the population of humans as a whole, each individual presents with a unique genotype. Even so-called identical twins have some genetic differences. Thus within an essentially genetically homologous group, the individual stands out. As humans we thus exist as individuals within a dynamic environment, our physical evolution speeded up dramatically by culture. What is a clear and significant factor in understanding human nature is that it is impossible to present a fixed model of it as has traditionally been done in many religious and philosophical theories. But we are not left hanging, so to speak, for we can reflect on our selves, our situations, and our experiences. We know that we are a species that engages in symbolic discourse and communicates efficiently and profoundly through language as no other organism can. Most importantly, we seek to find systems of meaning for our lives that help us to make sense of our experiences, as well as to transform them.

The shift to genetics and genetic understandings of both evolution and human origins, as well as the genome project itself, has given rise to a variety of explanations of human nature and behavior loosely grouped under the heading of sociobiology. In the next section of this paper, I will examine both general sociobiological claims as well as various philosophical perspectives that ground these claims, focusing thematically on freedom, altruism, and religion. I will engage in a discussion of these claims from other philosophical perspectives as a way to give insight into our elusive, dynamic, and changing human nature.

2. HUMAN NATURE IN THE CONTEXT OF MODERN BIOLOGY

In this section I will move from these general considerations about humans in relation to modern biology into some specific considerations of human nature. I wish to consider three specific questions that have historically been associated with human nature but that have been challenged or seen as irrelevant in light of modern biology. These are the questions of freedom, altruism, and a capacity for transcendence or religion.

These three characteristics of humans have typically been understood as qualities that separate us from other animals, give us a particular relation to our own actions and other beings, as well as provide a sense of meaning that transcends our biological fate. Contemporary commentators have also singled out these characteristics for analysis. In this section I wish to join this debate by incorporating aspects of other philosophical traditions as well as aspects of contemporary thought to help develop some insights into our human nature.

Sociobiological Perspectives

In this section I will present a sampling of sociobiological perspectives on two core problems historically associated with an understanding of human nature. I do this, first, to set a general context for our discussion, and then, second, to identify particular problems that can then be addressed in light of other philosophical perspectives.

Freedom

The question of freedom and determinism is an ancient philosophical question but it is also proving to be a critical scientific one as well. Knowledge of the action of specific genes as well as the action and interaction of hundreds of genes has focused on the question of freedom in a particularly sharp way. The discipline of sociobiology in particular has helped to refocus our attention on this question. A general problem in sociobiology is the tendency to assume that what is true of animal behavior is also true of human behavior. Thus one could assume that since a large part of the human genome is shared with other animals, we are simply following our genetic programs as they do. Some respond to this by noting the presence of culture, understood broadly in a social and biological sense, as a mediating force on our genome. Yet while this can put a bit more play into our actions, some would see culture determining our actions as much as our genes do. So we need to attend carefully to the question of whether or not there is direct evidence of a genetic or cultural foundation for a particular trait or behavior and to what degree that foundation determines that behavior.

Another part of the problem is definitional. For example, E. O. Wilson responded in the following way to the question of whether the fact that the brain is programmed by the genes destroys free will: "The biases in mental development are only biases; the *influence of the genes, even when very strong does not destroy free will.* In fact, the opposite is the case: by acting on culture through the epigenetic rules, *the genes create and sustain the capacity for conscious choice and decision.*"[21] This is a clear rejection of determinism and a good example of gene-environment interaction. But it also identifies freedom as choice. While that is a common understanding of freedom, we need to reflect on whether it is a fully adequate understanding of freedom. Other authors, however, seem to qualify freedom by stating that: "while they exercise free will in moment-by-moment choices, this faculty remains superficial and its value to the individual is largely illusory," and "Real freedom consists of choosing our masters by a procedure that allows us to master them."[22] This statement presents freedom as illusory and in fact we have this illusion of freedom only because we choose what will determine our actions. But the resolution is unsatisfactory. While we seem to make ourselves genetically, which results in our genes controlling our actions, we also seem to be choosing the particular genes that do the controlling. This correlates with Wilson's position, noted above, that while sometimes genes hold culture on a leash, interestingly enough, culture holds the genes on a leash. How human freedom would fit here is quite unclear, for a leash is still a leash and in this perspective it sets clear limits.

Wilson goes on to specify the nature of freedom in the following way:

> To the extent that the future of objects can be foretold by an intelligence which itself has a material basis, they are determined—but only within the conceptual world of the observing intelligence. And insofar as they can make decisions of their own accord—whether or not they are determined—they possess free will.[23]

Wilson uses the example of a bee. If we were to know all the properties of small animals—for example, a bee's nervous system, its behavioral characteristics, and its personal history—and if this information could be on a computer program, we could predict the bee's flight. To the circle of human observers watching the computer readout, the future of the bee is determined to some extent. But in her own "mind" the bee, who is isolated permanently from such human knowledge, will always have free will.[24] The same is true for humans, insofar as their behavior can be specified. However, because of the complexity of human behavior and technical limitations, and perhaps the capacity of intelligence in general, such specification and prediction of human behavior is practically impossible. Wilson concludes:

> Thus because of mathematical indeterminacy and the uncertainty principle, it may be a law of nature that no nervous system is capable of acquiring enough knowledge to significantly predict the future of any other intelligent system in detail. Nor can intelligent minds gain enough self- knowledge to know their own future, capture fate, and in this sense eliminate free will.[25]

For Wilson free will is either indeterminacy or unpredictability and is a function of a technical inability either to know all the variables or—should they be known—to program them in a meaningful way.

Richard Dawkins also argues that there is no clear relation between a particular trait's being under genetic control and the possibility of its modification. While this argues against a particular kind of genetic determinism and lack of freedom, the question of how such modification occurs still remains. Dawkins is also clear to state, particularly in *The Selfish Gene*, that he is arguing how things evolved, not how humans ought to act. He is not particularly interested in humans and human behavior, but rather animal behavior. So again one must be careful how one reads and parses his observations. But, having been tarred with the brush of genetic determinism in critiques of the first edition of the book, Dawkins is quite blunt in his rejection of it:

> It is perfectly possible to hold that genes exert a statistical influence on human behaviour while at the same time believing that this influence can be modified, overridden, or reversed by other influences. . . . We, that is our brains, are separate and independent enough from our genes to rebel against them.[26]

The interesting part of this sentence is the identification of the self with the brain. How one understands that will also suggest something about freedom and how it functions.

And we must also remember that Dawkins, like Wilson, states that we have the capacity to rebel against our replications, as he notes in *The Selfish Gene*. But again, one must seek for the foundation or basis of such a capacity. Is this a capacity found generally in all animals, the focus of his study, or is it unique to humans?

One important part of the argument of both these authors is that any discussion of freedom must occur within a context, a context that is both genetic and cultural. No one stands apart from such an environment and this environment must at least condition or qualify both our understanding of freedom as well as its exercise. But since both argue for some capacity to transcend one's genetic program, we need to look carefully for the basis of that capacity.

Altruism

Altruism is a word describing a noble tendency in humans: actions on behalf of another with little or no regard for one's self or one's interests. In the literature of sociobiology, however, it is near equivalent to a fighting word. Generally it refers to some form of behavior that promotes the fitness of another organism at the expense of its own fitness. On the one hand, this is a *behavioral* term that describes how natural selection occurs, not a description of *motives*. On the other hand, Wilson, for example, argued that altruism is the central problem of sociobiology. With this he also brought a "particular philosophical style: the coupling of scientific and moral notions" and thus looked for holistic explanations of behavior that led him occasionally to commit the naturalistic fallacy of describing moral norms from biological descriptions.[27] For Richard Dawkins, discussing this from the gene's perspective, the point is not the survival of the individual, but survival of copies of the genes. Since relatives are the ones who share these genes, altruistic behavior toward relatives is to be expected.

As if this were not problematic enough, the term also is involved in a dispute over the workings of natural selection with the debate falling roughly between group selection and kinship selection, with the term inclusive fitness also being introduced for good measure. Historically, most argued that natural selection proceeded through group selection, that is, through behavior that was to the advantage of the group. In this model, altruistic behavior was self-sacrificial behavior for the good of the group. The late William Hamilton developed a complex mathematical argument for kin selection. This was altruistic behavior on the part of the individual "towards relatives with whom they have genes in common."[28] Inclusive fitness, again a concept developed by Hamilton, "explains how natural selection can favor altruism. This can happen if the benefits of altruism can be made to fall on individuals who are likely to be altruist rather that random members of the population."[29] Thus from Hamilton's perspective, inclusive fitness is a broader term that can include both kin and group selection as mechanisms for the evolution of altruism.

Now the problem: is this explanation relevant to human behavior? Is this mechanism of natural selection operative in our nature as well? Are we genetically predisposed to favor our relatives over others? In a controversial paper, Hamilton argued the following:

> It can even be suggested that certain genes or traditions of the pastoralists revitalize the conquered people with an ingredient of progress which tends to

> die out in a large panmietic population for reasons already discussed. I have in mind altruism itself or the part of altruism which is perhaps better described as self-sacrificial daring. By the time of the renaissance, it may be that the mixing of genes and cultures (or cultures alone, if these are the only vehicles, which I doubt) has continued long enough to bring the old mercantile thoughtfulness and infused daring into conjunction in a few individuals who then find courage for all kind of inventive innovation against the resistance of established thought and practice. Often, however, the cost in fitness of such altruism and sublimated pugnacity to the individuals concerned is by no means metaphorical, and the benefits to fitness, such as they are, go to a mass of individuals whose genetic correlation with the innovator must be slight indeed. Thus civilization probably slowly reduces its altruism of all kinds, including the kinds needed for cultural creativity.[30]

This line of argumentation certainly suggests that when the benefits of an altruistic act do not go to relatives, the benefits tend to disappear over time. Additionally, the argument seems to suggest that acting against one's genes or against natural selection decreases the number of such genes in the population as well as decreases the overall fitness of society. Wilson phrases the issue this way:

> Can the cultural evolution of higher ethical values gain a direction and momentum of its own and completely replace genetic revolution? I think not. The genes hold culture on a leash. The leash is very long, but inevitably values will be constrained in accordance with their effects on the human gene pool. The brain is a product of evolution. Human behavior—like the deepest capacities for emotional response which drive and guide it—is the circuitous technique by which human genetic material has been and will be kept intact.[31]

For Wilson, the genetic program is key to understanding human development on all levels. Thus, while the culture may move in a particular direction, eventually and ultimately it will be conformed to the genetic program, and group and kin selection will win out.

Wilson makes altruism the central theoretical problem of sociobiology. This is so because in a "Darwinist sense the organism does not live for it self. Its primary function is not even to reproduce other organisms; it reproduces genes, and it serves as their temporary carrier."[32] This occurs through natural selection "a process whereby certain genes gain representation in the following generations superior to that of other genes located at the same chromosome positions."[33] Thus the organism is but DNA's way of making more DNA, and the individual but the vehicle of the genes.

In this context, the question is how can altruism—"self-destructive behavior performed for the benefit of others"[34]—possibly evolve through natural selection. Obviously this behavior reduces personal fitness and would seem to lead to the loss of the gene or genes responsible for that behavior. Wilson finds the answer to this question in kinship.

> [I]f the genes causing the altruism are shared by two organisms because of common descent, and if the altruistic act by one organism increases the joint contribution of these genes to the next generation, the propensity to altruism will spread through the gene pool. This occurs even though the altruist makes less of a solitary contribution to the gene pool as the price of its altruistic act.[35]

Wilson argues that "the impulse need not be ruled divine or otherwise transcendental, and we are justified in seeking a more convenient biological explanation."[36] Though Wilson notes that specific forms of altruism are culturally determined, he argues that the sociobiological hypothesis "can explain why human beings differ from other mammals and why, in one narrow aspect, they more closely resemble social insects."[37]

Wilson further distinguishes two forms of cooperative behavior. First is what he terms hard-core altruism: "the altruistic impulse can be irrational and unilaterally directed at others; the bestower expresses no desire for equal return and performs no unconscious actions leading to the same end."[38] Here the responses are unaffected by social reward or punishment and tend to serve the "altruist's closest relatives and to decline steeply in frequency and intensity as relations become more distant."[39]

Second is soft-core altruism: the altruist "expects reciprocation from society for himself or his closest relatives. His good behavior is calculating."[40] Thus soft-core altruism is essentially selfish in a traditionally moral sense as well as being influenced by cultural evolution. The psychological vehicles for this behavior are "lying, pretense, and deceit, including self-deceit, because the actor is most convincing who believes that his performance is real."[41]

In Wilson's perspective, soft-core altruism is the key to human society because it broke the constraints on the social contract imposed by kin selection. Reciprocity is the key to the formation of society. Hard-core altruism, on the other hand, is the "enemy of civilization."[42] This favors kin selection, the favoring of one's own relatives, and permits only limited global cooperation. Thus he says,

> Our societies are based on the mammalian plan: the individual strives for personal reproductive success foremost and that of his immediate kin secondarily; further grudging cooperation represents a compromise struck in order to enjoy the benefits of group membership.[43]

This gives Wilson a basis for optimism, for he thinks humans are "sufficiently selfish and calculating to be capable of indefinitely greater harmony and social homeostasis. This statement is not self-contradictory. True selfishness, if obedient to the other constraints of mammalian biology, is the key to a more nearly perfect social contract."[44]

> These other constraints are learning rules and emotional safeguards. Thus, honor and loyalty are reinforced while cheating, betrayal, and denial are universally rejected. Thus it seems that learning rules, based on innate, primary reinforcement, led human beings to acquire these values and not others with reference to members of their own group. . . . I will go further to speculate that the deep structure of altruistic behavior, based on learning rules and emotional safeguards, is rigid and universal. It generates a set of predictable group responses.[45]

Thus soft-core altruism provides the basis for various social allegiances, shifting though they may be. The critical distinction is the in-group and the out-group, the line between which fluctuates continually. But this is our social salvation for, if hard-core altruism were the basis of social relations, our fate would be a continuous "intrigue of nepotism and racism, and the future bleak beyond endurance."[46] Soft-core altruism provides an optimistic cynicism that can give us the basis of a social contract. Such behavior has been "genetically assimilated and is now part of the automatically guided process of mental development."[47] Thus genes hold culture on a leash and though the leash is long, "inevitably values will be constrained in accordance with their effects on the human gene pool."[48]

Richard Dawkins, who is most popularly associated with a narrow reading of altruism through the publicity and controversy surrounding *The Selfish Gene*, explicitly rejects any direct application of his explanation of evolution to human behavior. Two things work against him, however. First, he describes the evolutionary mechanism from the perspective of the gene and highlights the interest of the gene in producing replicas of itself rather than the individual as such. The choice of the metaphor of selfishness, rather than cooperation perhaps, suggests a motive rather than a behavior, a motive that could easily be applied to human behavior. Second, the preface to the first edition written by Richard Trivers, states:

> In short, Darwinian social theory gives us a glimpse of an underlying symmetry and logic in social relationships which, when more fully comprehended by ourselves, should revitalize our political understanding and provide the intellectual support for a science and medicine of psychology. In the process it should also give us a deeper understanding of the many roots of our suffering.[49]

In short, for those who wanted to read a theory of human behavior into *The Selfish Gene*, such an opportunity was handed them on a silver platter. Yet the question remains: are we on a genetic leash? Do we act to benefit primarily our relatives? Is action beyond the group possible or, as Hamilton suggested, will civilization gradually erode self-sacrificial behavior? This point is complicated by the absence of this preface in the 1989 edition, together with these sentences in the first chapter:

> My purpose is to examine the biology of selfishness and altruism. . . . Apart from its academic interest, the human importance of this subject is obvious. It touches every aspect of our social lives, our loving and hating, fighting and cooperating, giving and stealing, our greed and our generosity.[50]

Even though Dawkins affirms that his focus is behavior not motive—the effects of one's act, not one's subjective dispositions—the language here certainly is open to a discussion of motives, even though there is a strong attempt to redefine such terms. Thus, in the definition of altruism as behavior "to increase another such entity's welfare at the expense of its own,"[51] welfare is understood as one's chance of survival. One looks at outcome, not motives. Thus, a selfish gene tries "to get more numerous in the gene pool. Basically the gene does this by helping to program the bodies in which it finds itself to survive and to reproduce."[52] However—and this is a key issue for this section—"a gene might be able to assist *replicas* of itself which are sitting in other bodies. If so, this would appear as an act of individual altruism but it would be brought about by gene selfishness."[53]

The key way in which such genetically altruistic acts occur is through kin selection or within-family altruism, one that increases the greatest net benefit to one's genes, i.e., ensures the highest success rate for a particular gene. As Dawkins phrases it:

> A gene for suicidally saving five cousins would not become more numerous in the population, but a gene for saving five brothers or ten first cousins would. The minimum requirement for a suicidal altruistic gene to be successful is that it should save more than two siblings (or children or parents), or more than four half-siblings (or uncles, aunts, nephews, nieces, grandparents, grandchildren), or more than eight first cousins, etc. Such a gene, on average, tends to live on in the bodies of enough individuals saved by the altruist to compensate for the death of the altruist itself.[54]

Thus Dawkins concludes: "I have made the simplifying assumption that the individual animal works out what is best for his genes."[55]

This is essentially what Wilson calls hard-core altruism and described such behavior as "the enemy of civilization."[56] Soft-core altruism, recall, is what makes society possible, though to a limited degree only. Thus for Wilson, the

> most elaborate forms of social organization, despite their outward appearance, serve ultimately as the vehicles of individual welfare. Human altruism appears to be substantially hard-core when directed at closest relatives, although still to a much lesser degree than in the case of the social insects and

> the colonial invertebrates. The remainder of our altruism is essentially soft. The predicted result is a melange of ambivalence, deceit, and guilt that continuously troubles the individual mind.[57]

This perspective seems to leave us in a rather melancholy state at best and total despair at worst. From a biological perspective, both Wilson and Dawkins seem to have placed us squarely in the middle of a Hobbesian world. This view was promulgated most clearly in Dawkins' *The Selfish Gene,* the main argument of which was that "a predominant quality to be expected in a successful gene is ruthless selfishness. . . . Much as we might wish to believe otherwise, universal love and the welfare of the species as a whole are concepts which simply do not make evolutionary success."[58] Indeed as Dawkins says, "I think 'nature red in tooth and claw' sums up our modern understanding of natural selection admirably."[59]

Conclusion

So where do these considerations leave us in considering human nature? First, I think in a very confusing place. In part this is because terms and their meaning vary from author to author. But it is also because authors are attempting to develop integrating theories of human behavior without appealing to motives. Nonetheless there is an appeal to some kind of an ethical theory on which people can be held accountable. Second, the authors operate out of an evolutionary framework that shapes their perspectives. They correctly note that we simply cannot speak of human behavior without simultaneously speaking of genes and their effect on the total organism. But, third, the primacy seems to be on the role of the genes. Though the authors explicitly affirm the role and significance of culture, they return to the role of the gene. Wilson is most explicit when he says culture is held on a genetic leash. Dawkins is more ambiguous when he says we can rebel against our culture, but the basis for that is not clear. Additionally, one would wonder if there was a genetic consequence for straying from our genetic program. Also, connections between these perspectives and the HGP's tilt toward genetic essentialism are easy to make. The HGP will reinforce the search for the role and consequences of single genes. In turn, this will heighten the search for genetic programs that control our behavior.

The questions for examination are very complex. Are we simply matter, are we at the disposal of our genes, is there a basis for a kind of rebellion against our genes, is there a human nature, is freedom an illusion, do we have the capacity to transcend our nature? These questions press us from the perspective of the HGP as well as contemporary studies in genetics. In the following section I will begin addressing some of these questions from the perspective of the medieval philosopher John Duns Scotus, who I hasten to add is innocent of any knowledge of modern genetics or sociobiology. But Scotus appeals to aspects of our experience that I think are helpful in addressing the questions that have been raised. Minimally, he shows us another perspective from which to consider these issues; at the most he may provide a way to begin addressing the deep questions that must be raised about human nature in light of modern genetics.

3. PERSPECTIVES FROM THE PHILOSOPHY OF JOHN DUNS SCOTUS

The move from contemporary genetics to a medieval philosopher may seem strange or bizarre to some (or many). Yet I have become convinced that some of the ideas that Scotus developed in his writings can shed light on some aspects of our contemporary problem. Now, since Scotus died in 1302, it is obvious that he had neither knowledge of the theory of evolution nor any concept of what sociobiologists refer to as a reproductive strategy. Thus I am not attempting to bootleg any such theories into his thought. Nor will I use his ideas as a Procrustean bed with which to shape contemporary ideas. Rather, my sense is that Scotus has some insights that can help clarify the conundrum into which the sociobiologists seem to have gotten themselves. Thus I wish to focus in particular on his concepts of nature, freedom, and transcendence as a way to help think through some of the problems posed in the sociobiology debate.

Scotus's Concept of Nature

I argue that sociobiologists have made a major mistake in their use of the term altruism. My issue is the term, not the behavior—although my concern is not exclusively semantic. That is, while the behaviors described are biologically accurate—insofar as they stick to biology—the significance of these behaviors also has been misinterpreted primarily because of the sociobiologists' almost idiosyncratic use of the term altruism. And it is because of this that they have gotten themselves into what many consider to be a Hobbesian world.

Duns Scotus begins with two distinctions. First is the concept of a nature: a principle of activity by which an entity acts out or actualizes its reality. A being's nature is the reason why an entity acts as it does. Or as he says: "the potency of itself is determined to act, so that so far as itself is concerned, it cannot fail to act when not impeded from without."[60] A nature essentially explains why an entity acts as it does.

A will, on the other hand, "is not of itself so determined, but can perform either this act or its opposite, or can either act or not act at all."[61] Thus, the reason why this act was done as opposed to another is that the will is the will and can elicit an act in opposite ways. Following Anselm, Scotus distinguishes two movements in the will as the *affectio commodi*—the inclination to seek what is advantageous or good for one's self—and the *affectio justitiae*—the inclination to seek the good in itself.

In this section I focus on the *affectio commodi*, the will to do what is to our advantage, perfection, or welfare. This affection or inclination is a nature seeking its own fulfillment. For Scotus, this *affectio commodi* is not an elicited act. Rather it is a natural appetite necessarily seeking its own perfection. As Scotus says:

> That it does so necessarily is obvious, because a nature could not remain a nature without being inclined to its own perfection. Take away this inclination and you destroy the nature. But this natural appetite is nothing other than an inclination of this sort to its proper perfection; therefore the will as nature necessarily wills its perfection, which consists above all in happiness, and it desires such by its natural appetite.[62]

Allan B. Wolter provides an interesting commentary on this concept:

> All striving, all activity stems from an imperfection in the agent. As the etymological derivation of the word itself suggests, nature [from *nascor*, to be born] is literally what a thing was born to be, or more precisely, born to become, for nature as an active agent is essentially dynamic in a Faustian sense. It is restless until it achieves self-perfection. Since what perfects a thing is its good and since this striving for what is good is a form of love, we could say with Socrates that all activity is sparked by love.[63]

This love, however, is neither objective nor directed to the good of another, regardless of whether or not this other being might be a kin. It is self-centered and directed to seeking its own welfare. As Wolter further comments,

> If at times we encounter what seems to be altruistic behavior in the animal world, for instance, it is always a case where the "nature" or "species" is favored at the expense of the individual. But nature, either in its individual concretization or as a self-perpetuating species, must of necessity seek its own perfection. Such is its supreme value and the ultimate goal of all its loves.[64]

As Wolter interprets Scotus here, when an individual entity or a nature acts, it seeks its own good or what is to its advantage. This is not cause for surprise for this is what a nature does, whether looked at as an individual representative of the species or as the species as a whole. The *affectio commodi* drives the being "to seek his perfection and happiness in all he does."[65]

What is significant about this perspective—particularly in the context of the sociobiologists—is that for Scotus, and indeed for the entire classical philosophical tradition from Plato forward, seeking one's own perfection is a *good*. It is "not some evil to be eradicated. For it too represents a God-given drive implanted in man's rational nature which leads him to seek his true happiness."[66] In fact, to ignore our perfection or to give it no standing in our actions is an act of injustice to one's self.

I argue that what Wilson and Dawkins refer to as genetic selfishness is what Scotus refers to as the *affectio commodi*. The significance of the Scotistic position is, on the one hand, that he sees the same kind of tendency present in human nature as do the sociobiologists but, on the other hand, he, together with the entire philosophical tradition to that time, sees that behavior as a good because it achieves the perfection of the individual and the species. That is, for Scotus, the *affectio commodi* is that dimension of human nature that leads us to seek our fulfillment or perfection as a human. This affection is a good precisely because it leads to our perfection.

There is, however, a critical difference between Scotus and the sociobiologists. For the sociobiologists the behavior comes from evolutionary success, whereas for Scotus the cause is the creative will of God expressed in creation. Nonetheless, though the origin is quite different, the behavior is the same. Part of the difference surely lies in both philosophical and theological frameworks. But another part of the difference lies in that Scotus sees self-perfecting behavior as a good while the sociobiologists describe this as selfish—which even in their framework has a negative connotation.

But there remains this issue raised by the sociobiologists: is such genetically selfish activity the only possible mode of human activity? Or as Scotus would phrase it, can we see and actualize a good beyond ourselves and our perfection, beyond the *affectio commodi*? Scotistic thought would agree with the sociobiologists that as natures we, like any other nature, seek our good and our individual perfection and that we do so necessarily. But it would disagree that this is selfish in the pejorative sense of sociobiology. In fact, I think from a Scotistic perspective the sociobiologists' discussion of genetic selfishness makes no sense at all and is a significant distortion of human existence as the next section will argue.

Scotus on Freedom and Altruism

Scotus calls the *affectio justitiae*, or the affection for justice, the source of true freedom or liberty of the will and is the basis for his claim that true freedom goes beyond freedom understood as choice. Additionally, the *affectio justitiae* is the means by which we can transcend nature and go beyond our individually defined good and ourselves to see the value of another being.

> To want an act to be perfect so that by means of it one may better love some object for its own sake, is something that stems from the affection for justice, for whence I love something good in itself, thence I will something in *itself*.[67]

Allan B. Wolter notes four characteristics of the *affectio justitiae.* First, it gives us the capacity to love a being for itself rather than for what it can do for us. Second, it enables us to love God for who God is rather than for the consequence of God's love on us. Third, the *affectio justitiae* allows us to love our neighbor as ourselves thereby making each individual of equal value. Finally, such a seeking for the good in itself leads to a desire to have this good beloved by all, rather than being held to oneself.[68] This leads Wolter to the conclusion that the *affectio justitiae* amounts to a "freedom *from* nature and a freedom *for* values."[69] Or, as Scotus puts it:

> From the fact that it is able to temper or control the inclination for what is advantageous, it follows that it is obligated to do so in accordance with the rule of justice that it has received from a higher will.[70]

Such an understanding of will as *affectio justitiae* frees the will from the constraints of the necessity of human nature's act of self-realization or the seeking of its own good *only.* For Scotus then, when a free agent acts according to nature to realize itself or to seek its own good it paradoxically acts *unnaturally*, since to seek what is "*bonum in se* is not to seek something that 'realizes the potential of a rational nature.' It is somehow to transcend 'the natural' and thus to have a mode of operation that sets the rational agent apart from all other agencies."[71]

This understanding of will grounds, for Scotus, the possibility of our being able to transcend our own self-interest or self-benefit—what sociobiologists call genetic selfishness—a topic to be addressed later.

Scotus also proposes a view of freedom that is not limited to choice of alternatives or freely elicited acts. Rather, in keeping with his mentors Augustine and Anselm, Scotus views freedom as "a positive bias or inclination to love things objectively or as right reason dictates."[72] The proper focus of freedom, and by implication moral analysis, is not the individual act of choosing, but the inclination as a whole. And such an inclination focuses on fidelity to the good in itself, not the specific act of choosing that good nor the necessary appreciation of what is good for the fulfillment of the nature of the agent. Here Scotus follows the older Catholic tradition of Anselm, who said, "Whoever has what is appropriate and advantageous in such a way that it cannot be lost is freer than he who has this in such a way that it can be lost."[73] From a psychological point of view, Scotus argues that our awareness of the limitation of any particular act of will means that we experience freedom as choice. That is, we are aware that we could have chosen otherwise and that such a choice would have given a different degree of perfection. Thus, "choice is simply basic freedom in inferior conditions,"[74] i.e., human finitude. When we will or make a choice, our will is never fully actual or fully expressed, for it is contingent—we can in fact choose this option or that option. Yet for all that, we can approach our perfection through our steadfastness or constancy in cleaving to the object of our love. "The perfection of freedom connotes a perseverance and stability in the will's adherence to the good."[75]

Scotus presents both a critical and a positive perspective on freedom that is of particular importance. He discounts the significance of choice, understood as any particular choice or as any choice considered as an isolated event. To say this, of course, flies totally in the face of certainly the normative American experience of freedom and perhaps the Enlightenment tradition as well. For we revel in individual choice and assume that this is the essence of freedom. Such freedom is the core of autonomy, our expression of self-determination. From early in our lives, we are taught that ahead of us lies a series of decisions that will shape our lives and for which we alone will be responsible. As Americans, we have taken to heart existentialism's perspective that our existence precedes our essence and that one becomes one's self only through particular, individual choices. And, if such choices are absent, one remains unauthentic.

Scotus fashions his development of freedom "from above," from a theological perspective that grapples with the question of how God can be free if love for the divine essence—for only an infinite being can fulfill an infinite being—is necessary. Scotus develops two formulations of freedom to respond to this. The first looks to love for finite objects and is the ability *"not to limit* oneself to limitedly perfecting objects."[76] The second envisions love for God and freedom as the "ability to *continually adhere* to the unlimitedly perfecting object. The point common to both formulations is the will's ability to achieve perfection "through active union with its beloved."[77] This holds true regardless of whether the will is infinite and *de facto* there is no other intentional object or whether the will is finite and there are multiple intentional objects. Thus for Scotus the essence of freedom is not choice but what he calls *firmitas* or what we could call fidelity or constancy.

What follows from this is that the finite will can never fully express its basic freedom because for humans there will always be another intentional object, another "what if I would have done this?" that would lead to another version of myself. Thus for us to choose one goal is to abandon others together with the perfection they could have given us. And given that we are finite, we are not able—as is God—to choose that which would ultimately perfect us. Freedom thus manifests itself in choice: "basic freedom in inferior conditions,"[78] that is, in the context of finitude.

For Scotus, however, free will is not limited only to the fact of choice or even appropriately characterized by it. Rather, choice is "reflective of a deeper structure at work in a specific situation."[79] And this deeper structure is steadfastness which constitutes the perfection of the will: "a perseverance and stability in the will's adherence to the good."[80] And it is in this steadfastness of commitment that we attain our perfection, not in particular choices regardless of their relation to the good.

The affection for justice is the capacity to love something or someone for their own selves, regardless of whether this happens to be a good for me or not. As Wolter phrases it, this is a "freedom *from* nature and a freedom *for* values."[81] The conclusion is the paradox that

> what differentiates the will's perfection as nature from the perfection of all other natural agents is that it can never be attained if it be sought primarily or exclusively: only by using its freedom to transcend the demands of its nature, as it were, can the will satisfy completely its natural inclination.[82]

Scotus's affirmation here is that we have the capacity to value an entity for its own sake, independent of its personal or social utility. As Scotus would phrase it, we have the ability to transcend the capacity to do justice to ourselves by doing justice to the good itself. The strong claim is that we are capable of recognizing goods distinct from our self-perfection and independent of our interests and choosing them even though such a choice may run counter to our personal self-interest or what does justice to my own nature.

> The will by freely moderating these natural and necessary tendencies to happiness and self-perfection is able to transcend its nature and choose Being and Goodness for their own sake. . . . Thus the free will is not confined to objects or goods that perfect self, but is capable of an act of love. . . . [L]ove is the most free of all acts and the one that most perfectly expresses the will's freedom to determine itself as it pleases.[83]

The conclusion is that one can distinguish at least a good and a better in human life. What is good in human life is a life that perfects us, that brings our being to a greater actualization. This is the realization of the *affectio commodi*. But what is better is the transcendence of self either to appreciate goods independent of us or even to curb our legitimate interest in self perfection to seek the good of others for their own sake. This is the realization of the *affectio justitiae*.

Put existentially,

> A free choice, then is the meaning of existence and the total initiative is left to man to rightly moderate his natural tendencies in the pursuit of being for its own sake. And in this sense one's existence is one's own responsibility and depends on one's causal initiative as an ultimate response to Being or Nothingness.[84]

Put ethically,

> right reason also recognizes that our self-perfection, even through union with God in love, is not of supreme value. It enables man, in short, to recognize that the drive for self-perfection paradoxically must not go unbridled if

> it is to achieve its goal, but must be channeled lest it destroy the harmony of the universe intended by God.[85]

What is most helpful about this perspective is that while it affirms self-perfection, ultimately such perfection is not an end in itself. To "be all that we can be" we must step beyond the confines of self and actualize that most free of all acts, an act of love. For only then do we find ourselves open to the depths of reality. And in the steadfast adherence to that beloved, we realize the fullness of freedom.

4. RELIGION

In this section I explore the general topic of religion, particularly with respect to the idea of its very possibility. I propose to examine the foundations that could make such a reality possible. I also will examine the adequacy of the philosophy of scientific materialism in capturing the sufficiency of matter as well as attitudes about religion expressed by various authors.

Scientific Materialism

Although the term scientific materialism appears late in Wilson's *On Human Nature,* scientific materialism is a key principle that provides the overarching framework for many of the ideas in sociobiology. Scientific materialism is "the view that all phenomena in the universe, including the human mind, have a material basis, are subject to the same physical laws, and can be most deeply understood by scientific analysis."[86] The core of scientific materialism is the evolutionary epic whose minimum claims are

> that the laws of the physical sciences are consistent with those of the biological and social sciences and can be linked in chains of causal explanation; that life and mind have a physical basis; that the world as we know it has evolved from earlier worlds obedient to the same laws; and that the visible universe today is everywhere subject to these materialist explanations.[87]

Scientific materialism is a mythology and "the evolutionary epic is probably the best myth we will ever have," and it can be "adjusted until it comes as close to truth as the human mind is constructed to judge the truth."[88]

Of critical importance is a discussion of matter, the ultimate grounding—so to speak—of evolution. In Wilson's theory, matter is all that is and all that is needed to account for all activity—insect or animal, private or social. For Wilson, matter is most creatively expressed in the gene, the basic unit of heredity and "a portion of the giant DNA molecule that affects the development of any trait at the most elementary biochemical level."[89] Thus we need to examine human nature through biology and the social sciences. This will lead us to an understanding of the mind

> as an epiphenomenon of the neuronal machinery of the brain. That machinery is in turn the product of genetic evolution by natural selection acting on human populations for hundreds of thousands of years in their ancient environments.[90]

The Transcendent Potential of Matter

But is matter only matter, inert particles interacting according to the laws of physics and chemistry, or is there another level?

One traditional theory explaining the interaction of particles of matter such as electrons and positrons is hylosystemism which holds that "all bodies, or at least non-living bodies, are composed of elementary particles or hylons which are united to form a dynamic system or functional unit."[91] In this context, system refers to "a functional nature, possessing new powers."[92] When put into various combinations or when actualized under various conditions, these elementary particles form new systems educed from the matter and the properties of this new system and are

> not simply the arithmetical sum of the actual properties manifested by these hylons in isolation for the property of any given system such as the nucleus or the hydrogen atom . . . is rooted proximately in the new powers of the respective system, powers which, though ultimately reducible to the two or more hylons that function as essentially ordered causes, exist only virtually in the individual hylons.[93]

Consequently, the properties of individual particles seen in isolation can never tell us the full range of these particles when combined into a system. Therefore, within matter lies a range of possibilities that emerge or are actualized only when these particles are put into a system or when a previous system is restructured.

What are the implications of such a theory? Karl Rahner argues that we are the beings "in whom the basic tendency of matter to find itself in the spirit by self-transcendence arrives at the point where it definitely breaks through."[94] For Rahner, "matter develops out of its inner being in the direction of spirit."[95] This becoming, a becoming more rather than becoming other, must be "effected by what was there before and, on the other hand, must be the inner increase of being proper to the previously existing reality."[96] This notion of becoming more is a genuine self-transcendence, a "transcendence into what is substantially new, i.e., the leap to a higher *nature*."[97]

While Rahner does not argue that life, consciousness, matter, and spirit are identical, he does argue that such differences do not exclude development:

> In so far as the self-transcendence always remains present in the particular goal of its self-transcendence, and in so far as the higher order always embraces the lower as contained in it, it is clear that the lower always precedes the actual event of self-transcendence and prepares the way for it by the development of its own reality and order; it is clear that the lower always moves slowly towards the boundary line in its history which it then crosses in actual self-transcendence.[98]

For Rahner, then, the human is the "self-transcendence of living matter."[99] On the one hand, Rahner describes this as the cosmos becoming conscious of itself in the human. Yet on the other hand again, this self-transcendence of the cosmos reaches

> its final consummation only when the cosmos in the spiritual creature, its goal and its height, is not merely something set apart from its foundation—something created—by something which receives the ultimate self-communication of its ultimate ground itself, in that moment when this direct self-communication of God is given to the spiritual creature in what we—looking at the historical pattern of this self-communication—call grace and glory.[100]

Lindon Eaves and Lora Gross present another presentation of matter as the ground of new potentialities. They argue for a dynamic and holistic conception of matter that emphasizes the "unity of matter, life, and energy and understands nature as a profoundly complex, evolving system of intricately interdependent elements."[101] They suggest a vitality in matter that gives it depth and intensity, value, and the inclination toward organization.

Eaves and Gross operate from a biological and specifically genetic perspective that "seeks a new framework for its comprehension that does justice to all the so-called higher aspects of human consciousness in a phylogenetic and ontogenetic framework"[102] This perspective focuses on the mechanisms of inheritance, which "have within themselves the probability of presenting new transcendent possibilities for action within history."[103] Thus they argue that surprise is inherent in nature and then develop a view of nature itself as gracious and argue, similarly to Rahner, that "genetics provides *a basis for grace within the structure of life itself*."[104]

This position serves as the basis for a rejection of crude determinism, for "the material processes of life have produced a person who transcends all conventional definitions of personhood to the point where the term *freedom* is the best we have available."[105]

This gives rise to two consequences: first, "culture creates conditions for completion in community that would otherwise be impossible in a mere aggregation of individuals"[106]; second, "recognition that the conditions of life are such that the process that produces pain, in the sense of genetic disease, is also the process that maintains life in the cosmos."[107]

This second point is critical in that it highlights the value of genetic diversity and provides the ground for criticizing simplistic models of genetic waste, unfitness, and disease. Additionally, this point recognized a fundamental ambiguity in the nature of reality. Cancer is a result of the extremely rapid division and growth of cells, the very same process that allows life to continue. In the process of genetic recombination in sexual reproduction, copying errors sometimes occur that result in disease. Yet it is this very same process that allows reproduction to occur at all. These biological processes are the means through which life is transmitted from one generation to another, yet it is through these very same processes that life can be transformed in ways that are sometimes new and helpful and sometimes new and harmful.

A similar point emerges from a consideration of the multiplicity of forms and species.

> There are many forms which do not constitute a value or an advantage in the struggle of life; they are useless in this sense, and for that reason they are beautiful. Beauty is a factor that is not necessitated by lower needs, but is something that supposes the liberty of artistic creation.[108]

Considerations such as these regarding the chemical composition of life expressed in the wondrously complex DNA molecule cannot help but also push us in the direction of a radical reconsideration of the nature of matter from both a religious and scientific perspective. For example, the theologian Zachary Hayes, O.F.M., expresses it this way:

> the biblical tradition is a religious tradition that is convinced of the deep religious significance of the material world and of its profound potential for radical transformation into a form so different from its present form in space and time (i.e., the idea of the incarnation and the metaphor of resurrection as the final condition of "becoming flesh").[109]

This is an echo of the medieval theologian St. Bonaventure who said: "Again, the tendency that exists in matter is ordained toward rational principles. And there would be no perfect generation without the union of the rational soul with the material body."[110] Although expressed in what we would consider dualistic language, Bonaventure suggests that matter has within it the potential to transcend itself. John Paul II also articulates this in his address on evolution, in which he speaks of an ontological leap in which something profoundly different appears within the material reality out of which humans evolve.[111] Such discussions of necessity force us into a more critical dialogue with contemporary physics, particularly quantum mechanics with its take on the nature of matter. While such a discussion is beyond the scope of this paper, I recognize the necessity for such dialogue as articulated in this question by Hayes:

> Do we have a spiritual substance such as a 'soul,' or are soul functions such as consciousness, etc., really symptoms of chemical complexification of matter that is still in the process of moving to its final, fulfilling form?[112]

Whatever the outcome of such a debate, the view of matter and evolution suggested here is in the tradition of Augustine and his follower Bonaventure who saw history as a most beautiful song, a "*pulcherrimum carmen* which has been played by the divine Wisdom since the first organisms were called into existence, and of which our present forms are but one scene."[113] Or, as the *Book of Proverbs* says of wisdom:

> I was by his side, a master craftsman, delighting him day after day, ever at play in his presence, at play everywhere in his world. (8:30)

Specific Religious Issues

Religion is an interesting test case in an examination of human nature, for what one says about religion also reveals a commitment to a particular ideology and perhaps a methodology. A particular problem is a frequent assumption that a commitment to methodological reductionism also implies a commitment to metaphysical reductionism, which does not necessarily follow. Here, above all, one's prior commitments to particular positions need to be attended to and examined carefully.

For example, Richard Dawkins adopts an explicitly antireligion position. He sees religion as superstition or myth (understood as a false statement) whose purpose is to hide scientific truths from the unsuspecting or the naïve. Faith, Dawkins declares "is such a successful brainwasher in its own favour, especially a brainwasher of children that it is hard to break its hold."[114] And in addition to faith's being an arbitrary belief—otherwise one could give reasons for one's position—faith leads to fanaticism:

> But it is capable of driving people to such dangerous folly that faith seems to me to qualify as a kind of mental illness. It leads people to believe in whatever it is so strongly that in extreme cases they are prepared to kill and die for it without the need for further justification.[115]

Dawkins defines the idea of God as a meme (Dawkins' term for a cultural unit of replication) and part of the meme pool. Thus the meme God gains its survival in this pool through its appeal to our psychology: "It provides a superficially plausible answer to deep and troubling questions about existence."[116] Thus for Dawkins God exists, but only as a meme within the culture.

The Blind Watchmaker, in addition to being a sustained argument for the randomness of evolution, is also an explicit attack on the proof of God based on design in nature. Here he is "advocating Darwinism not only as a candle in the dark against pseudo scientific beliefs, but also as a direct substitute for personal religion."[117] For Dawkins, evolution has no purpose other than the survival of particular genes, and which ones survive cannot be predicted in advance. Thus there is no design in the process of evolution.

Dawkins also says the following:

> You scientists are very good at answering "How" questions. But you must admit you are powerless when it comes to "Why" questions. . . . Behind the question there is always an unspoken but never justified implication that since science is unable to answer "Why" questions there must be some other discipline that is qualified to answer them. This implication is, of course, quite illogical.[118]

The question of course is on what basis does Dawkins make such a claim? Is this on the basis of scientific methodology? If so, what is it? On what basis does one determine that the differentiation of why and how questions is illogical? Is this a prejudice resulting from a precommitment to a metaphysical reductionism? It is one thing to reject religion, for what ever reason; it is quite another to argue that the rejection of religion follows directly from the acceptance of a scientific or Darwinian perspective.

Recall also that Dawkins argues that humans alone among all other species have the capacity to rebel against our genes. The basis on which one might do this is not clearly spelled out. Dawkins recognizes that we do this—the practice of artificial contraception is one of the stock examples of such behavior—but the justification for this is not completely or satisfactorily explained. Here it seems that there is some ambiguity in the nature of reality that escapes a totally scientifically materialistic explanation.

E. O. Wilson comes at the religion question from quite another perspective. First, Wilson was raised as a Southern Baptist and underwent a conversion experience as a youth. But he later underwent another conversion experience, one to evolution and against his own religious upbringing. This led him, according to Segerstråle, to want to "prove the (Christian) theologians wrong. He wanted to make sure that there could not exist a separate realm of meaning and ethics which would allow the theologians to impose arbitrary moral codes that would lead to unnecessary human suffering."[119] Important here is the strong identification of religion and ethics, which is not necessarily the case, as well as the desire to show that religion was not a privileged locus of knowledge for right and wrong. Here Wilson seems quite close to Dawkins in adopting a position of metaphysical reductionism.

On the other hand, Wilson, unlike Dawkins, is sympathetic to the "Why" questions that humans ask. For he recognizes that humans have deep emotional needs that must be satisfied. Here Wilson argues that

> our metaphysical quest is an evolutionary one: religious belief can be seen as adaptive. The submission of humans to a perceived higher power, in the case of religion, derives from a more general tendency for submission behavior which has shown itself to be adaptive. By submitting to a stronger force, animals attain a stable situation.[120]

In other words, Wilson here used ethological insight to argue that we cannot eliminate our metaphysical quest—it is part of our nature.

For Wilson, the choice is between empiricism and transcendentalism, whether philosophical or religious. His own preference is the empiricist view because it is objective, i.e., scientific. It proceeds by "exploring the biological roots of moral behavior, and explaining their material origins and biases."[121] And ultimately, the evolutionary myth of origins will replace the religious one.

Yet, Wilson leans towards deism and states that there could exist a cosmological God whose existence could be proved by astrophysics. On the other hand, "a biological God, one who directs organic evolution and intervenes in human affairs . . . is increasingly contravened by biology and the brain sciences."[122] For all this, though, Wilson says we need our transcendental beliefs: "*We cannot live without them*. People need a sacred narrative. They must have a sense of larger purpose in one form or another, however intellectualized."[123] However, a transcendental form of this narrative will not and cannot endure, for it eventually will not withstand scientific scrutiny. Thus our guiding narrative will need to be taken from "the material history of the universe and the human species."[124] But that is not to our or religion's disadvantage.

> The true evolutionary epic, retold as poetry, is as intrinsically ennobling as any religious epic. Material reality discovered by science already possesses more content and grandeur than all religious cosmologies combined. The continuity of the human line has been traced through a period of deep history a thousand times older than that conceived by the Western religions. Its study has brought new revelations of great moral importance. It has made us realize that Homo sapiens is far more than a congeries of tribes and races. We are a single gene pool from which individuals are drawn in each generation and into which they are dissolved the next generation, forever united as a species by heritage and a common future. Such are the conceptions, based on fact, from which new intimations of immortality can be drawn and a new mythos evolved.[125]

So, although disagreeing with Dawkins about the need of raising "Why" questions, Wilson essentially lands in the same place: metaphysical reductionism and materialism, for science ultimately will answer all questions. And the answer to the question "Why religion?" is that it is adaptive and leads to social stability.

In a concluding perspective on this, Wilson said in an interview

> I just believe, to put it as simply as possible, that science should be able to go in a relatively few decades to the point of producing a humanoid robot which would walk through that door. The first robot would think and talk like a Southern Baptist minister, and the second robot would talk like John Rawls. In other words, somehow I believe that we can reconstitute, recreate, the most mysterious features of human mental activity. That's an article of faith but it has to do with expansionism. That's expansionism![126]

Thus every element of mental and physical behavior will have a physical basis, and ultimately there will be a materialistic explanation for everything. For science will continue to test every religious assumption and claim about God and humans and ultimately will come to the foundation of all human moral and religious sentiments. "The eventual result of the competition between the two world views, I believe, will be the secularization of the human epic and of religion itself."[127]

> One would then test, in the sociobiological mode, whether the peculiarities of the human brain are inferred to have taken place. If such matching does exist, then the mind harbors a species god, which can be parsimoniously explained as a biological adaptation instead of an independent, transbiological force.[128]

Thus God, and religion, is a product of a brain that is a product of evolution, which leads us to various adaptive behaviors, of which religion is one. And we are back again to biology's being the full explanation of all behavior, Wilson's original point in developing his theory of sociobiology. But is this the whole story?

A problem here might be the lack of distinction between three types of "why" questions. A scientific why question seeks to answer how one could account for a particular outcome: why do bodies fall, for example. A philosophical question tries to seek out inner relations and ultimate principles—Aristotle's seeking out of final causes, for example. Religion pursues its why question in terms of ultimate meaning—for what may we hope, for example. Each of these disciplines has a particular set of rules and a framework in which its particular why question can be answered along with a set of criteria for evaluating the adequacy of the answer. A problem arises when one asks the why question of one discipline from within the perspective of another. Or when one insists on the criteria from one discipline as being the only criteria acceptable for verification. While it is the case that the boundaries of these disciplines more frequently resemble semipermeable membranes rather than fixed borders, one continually needs to be sensitive to what kind of question one is asking and what are acceptable criteria for evaluating an answer. Border crossings are to be expected in our interdisciplinary world, but one must also remember to respect the customs and culture of the territory we visit.[129]

5. CONCLUSIONS

Richard Dawkins is totally transparent in his disdain for religion. At best it is a holdover from a past filled with ignorance. At worst it is false security for the desperate and immature. Wilson recognizes the need for mythology and grandeur in human life. And he says this need will be admirably fulfilled by our evolutionary myth, the grandest myth we have. Yet he too, like Dawkins, winds up with a form of philosophical materialism that admits no transcendence, no reality other than matter.

Dawkins argues against a form of genetic predestination or determinism. He states that we are the only creatures who can rebel against our genes. Wilson too argues for a type of distance from the genes in that we can build various cultures, though he adds that the genes will always keep the culture on a leash of varying length.

We have here two central philosophical claims that have specific applicability to human nature: materialism and freedom. The claims serve as working hypotheses of the analytic framework for both of these men but are not given any full examination or defense. In both, the claims of philosophical materialism are strongly made, but yet both seem to want some slack cut for their conclusions. We continue to live in our world of "as if": as if we were free, as if belief made a difference, as if meaning mattered. In cold, hard reality, however, none of this may be true. For evolution is without direction, matter began and it will end, species evolve and go extinct, the world eventually ends. If anything, the human species is cursed because through consciousness it sees this and knows reality's inherent meaninglessness. Humans create myths, but they are groundless, adult "just so" stories to hide the impersonal march of natural selection, which is ultimately indifferent to anything.

"Science tells us that we are creatures of accident clinging to a ball of mud hurling aimlessly through space. This is not a notion to warm hearts or rouse multitudes."[130] This interesting rephrasing of Wilson's noble myth of evolution helps put one's finger on the nub of the problem: if rational explanations such as quantum physics and evolution are fully adequate explanations of our origins and our reality, why do we continue to read, create, and reformulate myths? Why have not the Gilgamesh epic, Beowulf, Exodus, the Bhagavad Gita, Pilgrim's Progress, the American Dream all vanished? Why has religion shown such a dramatic rebound following the break-up of the Soviet Union? Why is there such resistance to any form of transcendence in China?

It is facile simply to allege that this is clear and indisputable evidence for the reality and truth of religion. Yet the counterclaim that scientific explanations of the realities of birth, death, and tragedy suffice are equally facile. Yet that quote from Ehrlich points to an interesting opening or way to think about transcendence and freedom.

In his recent book *Human Natures*, Ehrlich refers to a theory developed by Jared Diamond called the Great Leap Forward, referring specifically to a shift in toolmaking that occurred at the end of the age of the Neanderthals.

> The change to that Upper Paleolithic technology which appeared first in the Middle East about 50,000–40,000 years ago, was the start of the most rapid and radical cultural change ever recorded in the hominid line . . . It is a leap into new technologies, art, and population growth—perhaps even into a new mode of speaking.[131]

What is interesting, according to Ehrlich, is that at a certain point, while brain size remained fairly constant, "cultural changes took place at astonishing speeds with no significant change in the physical appearance of people or in the characteristics of their brains that can be divined from fossil skulls."[132] The question, then, is "did the physical evolution of our ancestors' brains cause the Great Leap Forward—or did only the 'software' of culture change, not the 'wetware' of brain structure?"[133]

Here I want to reengage several of the ideas introduced earlier as a way of thinking about a basis for this Great Leap Forward and to suggest an alternative reading of the interpretations of Dawkins and Wilson.

In their discussion of evolution, Eaves and Gross focused on a view of matter, evolution, and genetics that sought to do justice to the reality of human consciousness. Such a focus looked to the mechanisms of inheritance that, in their words "have within themselves the probability of presenting new transcendent possibilities for action within history."[134] They then argued that such mechanisms are the basis for surprise and graciousness within nature and thus they feel confident in arguing that "genetics provided *a basis for grace within the structure of life itself*."[135]

This type of position was also articulated by Rahner who argued

> it is clear that the lower always precedes the actual event of self-transcendence and prepares the way for it by the development of its own reality and order; it is clear that the lower always moves slowly towards the boundary line in its history which it often crosses in actual self-transcendence.[136]

This echoes a much earlier tradition expressed in the writings of Bonaventure, the Franciscan theologian of the thirteenth century. In his Second Book of the Sentences, Bonaventure phrases the insight this way: "Thus nature, according to the Philosopher, always desires what is better; matter, which is composed of elementary forms, desires to be under mixed forms and that which is under mixed forms desires to be under complex forms."[137] Or as Bonaventure phrases it in his work *On Retracing the Arts to Theology*, "Again, the tendency that exists in matter is ordained toward rational principles, and there would be no perfect generation without the union of the rational soul with the material body."[138] While this language is clearly dualistic, it also takes seriously the reality of matter and expresses, in the language of medieval philosophy, a dynamic that is present within matter, a dynamic that carries matter beyond itself to a point of transcendence. Matter is not content to be itself but strives for ever greater complexity.

The upshot of all this, in the contemporary words of Eaves and Gross, is that "the material processes of life have produced a person who transcends all conventional definitions of personhood to the point where the term *freedom* is the best we have available."[139] They then note, as cited earlier, that culture produces conditions for community that are not possible in aggregates and that the very same conditions of life that produce tragedy also maintain life itself.

One example of such conditions could be the previously discussed Great Leap Forward in culture. But importantly this was followed by the Great Axial Age that lasted from about 800 to 400 B.C.E. This was the age that saw the rise of the great religions of the East, particularly Hinduism, Buddhism, and Confucianism. The West saw the rise of Zoroastrianism and Judaism. In Latin America we had the religions of the Aztec and Mayan civilizations. Additionally, in Greece the first philosophical speculations were being advanced. Although the Axial Age lasted but a few centuries, much was compressed into it—the foundations of classical religions and philosophy. And as in the Great Leap Forward of culture that preceded it, the question is why. *Homo sapiens* had been present for several centuries and civilizations had begun to flourish. But here was a new development—a focus on the transcendent, the other, the metaphysical, a world other than this one, but for many a world no less real than the one available to our senses.

One suggestion of course is that such efforts were but the first, feeble steps of what would become scientific explanations of the mysteries of the world. Another was that such speculations helped to shelter people from the terrors of nature or the fickleness of chance. And indeed in many cases this is probably a reasonable explanation. But the problem still remains—the same kinds of concerns, speculations, and searchings arising relatively simultaneously in separate geographical areas.

In these perspectives we have another development of the evolutionary process that gives a foundation for both of these culture-altering events as well as a response to the materialism of Dawkins and Wilson. The key point is that both our experience and our very culture point to another dimension of life, another quality that helps explain our drive for mythmaking, our drive to transcendence. Neither the Great Leap Forward nor the Great Axial Age can be dismissed. The point of contention is its basis. Complexity brought to a higher level is certainly one valid interpretation. But another dimension that has to be incorporated is the reality of genuine difference. Ehrlich notes this by observing that we share genes, but not cultures, with chimpanzees.[140] But we then create a culture that grounds the further creations of art, music, philosophy, and religion. Clearly the larger brain and its enormous complexity provide the biological substrate necessary for such a leap. But the continuing question of the brain is: is it necessary or necessary but not sufficient? The alternative reading that I suggest argues that such capacities are not added from without but are, in fact, the supreme fulfillment of matter manifested not only in a Great Leap Forward but also in a moment of self-transcendence such as those expressed in the Great Axial Age as well as in moments of individual transcendence.

One element in this is Ehrlich's rather straightforward admission that even though he is not a mind-matter dualist, his version of human nature "finds a strictly materialistic interpretation of the world unsatisfying."[141] While neither denying a form of materialism influenced by quantum physics nor the value of methodological reductionism, Ehrlich concludes, "We seem to be always forced back to the larger view to find a degree of satisfaction not provided by dissection of a problem into its smallest parts."[142] While this leads Ehrlich to conclude to a kind of practical dualism, the problem is still there: the parts do not adequately explain the whole.

The examination of freedom from Scotus's perspective also forces us to look beyond materialism, but without denying our own biological nature. Here we can reexamine Scotus's distinction between the *affectio commodi* and the *affectio justitiae* with complementary insights from Ehrlich. The *affectio commodi* states that a given nature will seek its own good. What this good is will be understood through an examination of this nature. The concept is open, in my judgment, to being understood in light of our knowledge of the nature of a particular organism in light of the best of our interdisciplinary or multidisciplinary knowledge. This would include, for example with human nature, how both biological and cultural evolution shape who we are, that is, how they define our nature. To say we act according to our nature is to say we act as we do because we have evolved into beings of a particular kind. Ehrlich refers to the experience of values that are connected to such direct feelings as perceived values. These are the immediate motivations or goods that guide our daily lives and actions and are tied closely to our evolutionary past.

> Whereas the motivation to get our genes into the next generation may be the distant cause of much of our behavior the immediate motivations are more

> familiar. We rarely mate to reproduce ourselves; we ordinarily mate because it feels good. We don't dodge an approaching car to preserve our ability to raise our children; we do it to avoid anticipated pain or death. We don't eat to gain energy; we eat to assuage hunger or for pleasure.[143]

This is about as clear a restatement of Scotus's *affectio commodi* as one would want. It affirms both the biological and cultural dimension of the formation of our nature, but it also avoids a genetic reductionism by not suggesting a gene for each action. We act as we do because of who we have become.

But for Scotus, this is not the end of the story, for we also experience another dimension to ourselves. In addition to the experience of pursuing our good to fulfill or complete ourselves, we also experience the desire to seek the good of another. This is Scotus's *affectio justitiae*, a check on our nature, if you will. This desire moves us in a different direction, not contrary to our nature, but transcending it. This affection leads to a pursuit of the good, in Scotus's perspective, for its own sake or to a pursuit of the good of one's neighbor. The *affectio justitiae* initiates a basic move beyond the parameters of our own nature to the situation of another.

Ehrlich calls such an experience empathy and relates this to the development of what he calls conceived values, values evolved to help deal with the social environment. This is also the arena of ethics, an evolved system of culturally shared understanding of right and wrong. Critical here is altruism, which sociobiology explains on the basis of inclusive fitness (for one's relatives) or reciprocity (for strangers). Ehrlich makes two interesting observations in relation to this. First, the origin of ethics cannot be traced to chimpanzees. "Chimps have no way to share values; ethics had to await at least the evolution of language, of an efficient method of sharing the ideas that were presumably generated by notions of empathy. There appears to be an unbridgeable gap between the ethical capabilities of human beings and those of chimpanzees."[144] Second, "empathy and altruism often exist where the chances for any return to the altruist are nil. Indeed, careful psychological experiments suggest that much of human helping behavior is divorced from any real prospect of reproductive or other reward.[145]

The basis of such behavior, Ehrlich argues, is empathy and "would seem a necessary prerequisite for such altruism, and many of our empathetic feelings are unrelated to personal advantage."[146] Empathy is an evolved capacity to feel for others and, while it will have a high degree of variability in its expression and may indeed have some limits to its expression, its presence is another fact of our experience not satisfactorily explained by appeal to our genome. What Ehrlich calls empathy is at least analogous to Scotus's concept of *affectio justitiae.*

The point of differentiation, of course, is the source of such an affection. For Ehrlich, empathy comes from the process of gene-culture coevolution.

For Scotus, the *affectio justitiae* is ultimately a result of our being created in a certain way by God, though Scotus is not a literalist in his understanding of how that creation occurred. But the more critical point, in my judgment, is that both have identified an extremely similar behavior in humans based on experience. Humans in fact can transcend their nature by stepping beyond themselves and acting for the benefit of another. Both affirm that humans have the capacity to see a good outside of themselves and to pursue it or use it as the basis for constructing an ethic. And here is the foundation that Dawkins needs to ground his claim that humans alone can rebel against their genes.

The presence of the *affectio commodi* and the *affectio justitiae* in the same person gives rise to a paradox, one noted earlier but now will be developed a bit more. The presence of both affections makes morality possible, but Scotus "cannot accept a theory such as Aristotle's where the moral is analyzed in terms of self-perfection or self-realization, i.e., in terms of the rational agent's inclination to realize the perfection of its nature."[147] That is, a morality based exclusively on the good of the individual agent cannot be the whole story, for it is a morality limited to one's own human nature. To pursue the *affectio justitiae* or to experience empathy is to transcend one's human nature and in a paradoxical way to act against one's nature. Thus in a free act—though not free in any unbounded or totally arbitrary sense—the agent can seek the good of another, as opposed to seeking one's genetic advantage only. I raise this issue not only to engage in a discussion of a critical experience on which we can begin to construct an ethic, but also to raise the question of whether such an experience of the *affectio justitiae* might also be a grounding of another transcendent experience—a religious experience. Can the experience of a good beyond one's self lead to an experience of some yet higher or perhaps ultimate good? Can such an experience be a transcendent point of opening for an encounter with the presence of an other or another dimension of reality? Reality seems to be open enough for such a reading, particularly when we recall Ehrlich's own dissatisfaction with a strictly materialistic reading of human experience. An expression of such a reflection on this possibility is the poem "God's Grandeur" by the Jesuit poet Gerard Manley Hopkins.

> The world is charged with the grandeur of God.
> It will flame out, like shining from shook foil;
> It gathers to a greatness, like the ooze of oil
> Crushed. Why do men then now not reck his rod?
> Generations have trod, have trod, have trod;
> And all is seared with trade; bleared, smeared with toil;
> And wears man's smudge and shares man's smell: the soil
> Is bare now, nor can foot feel, being shod.

And for all this, nature is never spent;
There lives the dearest freshness deep down things;
And though the last lights off the black West went
Oh, morning, at the brown brink eastward, springs
Because the Holy Ghost over the bent
World broods with warm breast and with ah! bright wings.[148]

NOTES

1. Etienne Fouilloux, "The Anteprepratory Phase: The Slow Emergence from Inertia: January 1959–October 1962," in *History of Vatican II*, vol. I, Giuseppe Alberigo, trans. Joseph A. Komonchak (Maryknoll, N.Y.: Orbis Books, 1995), 85–87.
2. Ullica Segerstråle, *Defenders of the Truth: The Battle for Science in the Sociology Debate and Beyond* (New York: Oxford University Press, 2000), 30.
3. Segerstråle, 307–08.
4. E. O. Wilson, *Consilience: The Unity of Knowledge* (New York: Alfred Knopf, 1998), 125ff.
5. Jonathan Marks, "98% Alike? What Our Similarity to Apes Tells Us about Our Understanding of Genetics," *The Chronicle of Higher Education*, 12 May 2000, B7.
6. Marks, 2000, B7.
7. Segerstråle, 94. Italics in original.
8. Richard Dawkins, *The Selfish Gene* (New York: Oxford University Press, 1989), 201.
9. L. Cavalli-Sforza, *Genes, People, and Language*, trans. Mark Seilstad (New York: North Park Press, 2000), 13.
10. Cavalli-Sforza, 11–12.
11. Natalie Angier, "Do Races Really Differ? Not Really, Genes Show," *New York Times*, 22 August 2000, F6.
12. Cavalli-Sforza, 23.
13. Cavalli-Sforza, 27.
14. Cavalli-Sforza, 78.
15. Cavalli-Sforza, 79.
16. Cavalli-Sforza, 81.
17. Andrew Watson, "A New Breed of High-Tech Detectives," *Science* 289 (11 August 2000): 851.
18. Cavalli-Sforza, 29.
19. *Ordinatio*, II, d3, p1, q7, a251.
20. Cavalli-Sforza, 65.
21. Segerstråle, 395.
22. Segerstråle, 398.
23. E. O. Wilson, *On Human Nature* (Cambridge: Harvard University Press, 1978).
24. Wilson, 1978, 73.
25. Wilson, 1978, 73–74.
26. Cited, Segerstråle, 395.
27. Segerstråle, 37.
28. Cited, Segerstråle, 472.

29. Cited, Segerstråle, 472.
30. Cited, Segerstråle, 47.
31. Cited, Segerstråle, 39.
32. E. O. Wilson, *Sociobiology: The Abridged Edition* (Cambridge: The Belnap Press of Harvard University, 1980), 3.
33. Wilson, 1980, 3.
34. Wilson, 1978, 213.
35. Wilson, 1980, 3.
36. Wilson, 1978, 152.
37. Wilson, 1978, 153.
38. Wilson, 1978, 155.
39. Wilson, 1978, 155.
40. Wilson, 1978, 155–56.
41. Wilson, 1978, 156.
42. Wilson, 1978, 181, 157.
43. Wilson, 1978, 199.
44. Wilson, 1978, 157.
45. Wilson, 1978, 162–63.
46. Wilson, 1978, 164.
47. Wilson, 1978, 167.
48. Wilson, 1978, 167.
49. Cited, Segerstråle, 78.
50. Dawkins, 1989, 1–2.
51. Richard Dawkins, *The Selfish Gene* (New York: Oxford University Press, 1976), 61.
52. Dawkins, 1976, 95.
53. Dawkins, 1976, 95.
54. Dawkins, 1976, 100.
55. Dawkins, 1976, 105.
56. Wilson, 1978, 157.
57. Wilson, 1978, 159.
58. Dawkins, 1976, 2–3.
59. Dawkins, 1989, 2.
60. *Questiones in Metaphysicam 1*, q. 15, A. 2 (Allan Wolter, O.F.M., ed., *Duns Scotus on the Will and Morality* (Washington, D.C.: The Catholic University of America Press, 1986), 151.
61. Ibid.
62. *Ordinatio* IV, suppl. dist., 49, qq. 9–10, in Wolter 1986, 185.
63. Allan Wolter, O.F.M., "Native Freedom of the Will as a Key to the Ethics of Scotus," in *The Philosophical Theology of John Duns Scotus*, edited by Marilyn McCord Adams (Ithaca, N.Y.: Cornell University Press, 1990), 150.
64. Wolter, 1990, 150.
65. Wolter, 1990, 151.
66. Wolter, 1990, 151.
67. *Ordinatio* IV, suppl. dist., 49, qq. 9–10, in Wolter, 1986, 185.
68. Wolter, 1990, 151.
69. Wolter, 1990, 152.
70. *Reportatio Parisiensis 11*, d. 6, q. 2, n. 9, in Wolter, 1990, 152.

71. John Bowler, "The Moral Psychology of Duns Scotus: Some Preliminary Questions," *Franciscan Studies* 28 (1990): 31–56.
72. Wolter, 1990, 152.
73. Felix Alluntis, O.F.M., and Allan B. Wolter, O.F.M., trans., *John Duns Scotus: God and Creatures* (Princeton University Press, 1975), 378.
74. William Frank, "Duns Scotus' Concept of Willing Freely: What Divine Freedom Beyond Choice Teaches Us," *Franciscan Studies*, 42 (1982): 68–89, at 87.
75. Frank, "Duns Scotus' Concept of Willing Freely," 98.
76. Frank, "Duns Scotus' Concept of Willing Freely," 83.
77. Frank, "Duns Scotus' Concept of Willing Freely," 83.
78. Frank, "Duns Scotus' Concept of Willing Freely," 87.
79. Frank, "Duns Scotus' Concept of Willing Freely," 85.
80. William Frank, "Duns Scotus' Quodlibetal Teaching on the Will" (Ph.D. dissertation, Catholic University of America, 1982), 77.
81. Wolter, 1990, 152.
82. Wolter, 1990, 154.
83. Valerius Messerich, O.F.M., "The Awareness of Causal Initiative and Existential Responsibility in the Thought of Duns Scotus," in *De Doctrina Ionnis Scoti*, vol. II. Problema Philosophica, Acta Congressus Scotistici Internationalis. Roma, 1968, 629–44, 630–631.
84. Messerich, 631.
85. Wolter, 1990, 153.
86. Wilson, 1978, 221.
87. Wilson, 1978, 201.
88. Wilson, 1978, 201.
89. Wilson, 1978, 216.
90. Wilson, 1978, 195.
91. Allan Wolter, O.F.M., *Select Problems in the Philosophy of Nature. Pro Manuscripto* (Olean, N.Y.: St. Bonaventure University, The Franciscan Institute, 1962), 98.
92. Wolter, 1962, 105.
93. Wolter, 1962, 105.
94. Karl Rahner, "Christology within an Evolutionary View," in *Theological Investigations*, vol. 5, trans. Kark-H. Kruger, (New York: Crossroad, 1983), 157–92, at 160.
95. Rahner, 1983, 164.
96. Rahner, 1983, 164.
97. Rahner, 1983, 165.
98. Rahner, 1983, 167.
99. Rahner, 1983, 168.
100. Rahner, 1983, 171.
101. Lindon Eaves and Lora Gross, "Exploring the Concept of Spirit as a Model for the God-World Relation in the Age of Genetics," *Zygon* 27 (1992): 261–85, at 226.
102. Eaves and Gross, 274.
103. Eaves and Gross, 278.
104. Eaves and Gross, 274.
105. Eaves and Gross, 275.
106. Eaves and Gross, 277.
107. Eaves and Gross, 278.
108. Eaves and Gross, 157.

109. Zachary Hayes, O.F.M., private correspondence, 15 January 2001.

110. St. Bonaventure, "On the Retracing of the Arts to Theology," in *The Works of St. Bonaventure*, edited by Jose De Vinck (Paterson, N.J.: St. Anthony Guild, 1966, vol. II, paragraph 20).

111. John Paul II, Message to Pontifical Academy of Sciences, 22 October 1996, para. 6.

112. Hayes, private correspondence.

113. Philotheus Boehner, O.F.M., "The Teaching of the Sciences in Catholic Colleges," Franciscan Educational Conference, 1955, 150–159, at 157.

114. Dawkins, 330.

115. Dawkins, 330.

116. Dawkins, 193.

117. Segerstråle, 400.

118. Cited, Segerstråle, 401.

119. Segerstråle, 38.

120. Cited, Segerstråle, 402.

121. Wilson, *Consilience*, 240.

122. Wilson, *Consilience*, 241.

123. Wilson, *Consilience*, 264. Italics in original.

124. Wilson, *Consilience*, 265.

125. Wilson, *Consilience*, 265.

126. Cited, Segerstråle, 160.

127. Wilson, *Consilience*, 265.

128. Cited, Segerstråle, 160.

129. Zachary Hayes, O.F.M., *A Window to the Divine* (Quincy, Ill.: Franciscan Press, 1999), 11–13.

130. Paul R. Ehrlich, *Human Natures: Genes, Cultures and the Human Prospect* (Washington, D.C.: Island Press, 2000), 214.

131. Ehrlich, 104.

132. Ehrlich, 106.

133. Ehrlich, 107.

134. Eaves and Gross, 278.

135. Eaves and Gross, 274.

136. Rahner, 1983, 167.

137. Alexander Schaefer, O.F.M., "The Position and Function of Man in the Created World According to St. Bonaventure," *Franciscan Studies*, 20 (1960): 318. My translation.

138. St. Bonaventure: "On the Retracing of the Arts," paragraph 20.

139. Eaves and Gross, 275.

140. Eaves and Gross, 204.

141. Ehrlich, 317.

142. Ehrlich, 318.

143. Ehrlich, 310.

144. Ehrlich, 311.

145. Ehrlich, 312.

146. Ehrlich, 313.

147. Bowler, 4.

148. Gardner, 1953, 27.

6

Human Gene Transfer: Some Theological Contributions to the Ethical Debate

James J. Walter

The international scientific community is in the midst of completing one of the most extraordinary endeavors in the history of science. It is seeking to discover the "Holy Grail" of our biological heritage by mapping and sequencing the entire human genetic code through the Human Genome Project (HGP). The human genome (i.e., all our genetic material) contains 30,000 to 40,000 genes (or possibly as few as 26,000)[1] that are located along our forty-six chromosomes. The chromosomes in the nucleus of our cells are made mostly from a DNA molecule whose structure is a double helix, and our genes, which consist of specific sequences of nitrogenous base pairs of adenine-thymine and cytosine-guanine, are located along this molecule. The smallest gene is comprised of a sequence of about 1,000 nitrogenous base pairs, and the largest gene has approximately two million. As one might imagine, mapping and sequencing this number of genes is a time-consuming enterprise. In June 2000, approximately 90 percent of the genome was mapped and sequenced, and now the task is to complete the remaining 10 percent. Francis Collins, the current director of the HGP, has stated that the public project should have a highly accurate map of the human genome completed well before 2003.[2]

James Watson, the former director of the U.S. component of this project, recognized early on that there are many important issues of a nonscientific nature connected with the genome initiative. He urged that, of the $3 billion (or $1 per base pair) that will be funded for the U.S. portion of the genome project, at least 3 percent ($90 million) should be spent on examining these issues. He succeeded in his efforts, and the Joint Working Group on the Ethical, Legal and Social Issues Relative to Mapping and Sequencing the Human Genome (ELSI) was formed and began its work in September 1989.[3] Watson was indeed correct about the relevance of the ethical issues connected to this initiative. The scientific breakthroughs that are being made today because of this research, and those that will be made in the future based on the various types of human gene transfer, present us with extremely complex and far-reaching theological, social, and moral questions.

My interest in the HGP is principally theological in nature and scope, although it involves many ethical issues as well. I will begin with a claim: none of us enters into the moral evaluation of a complex topic such as this one as if one were a *tabula rasa*, or empty slate. Rather, our moral judgments are informed and guided by settled convictions and beliefs of a nonmoral nature. For religious believers, these convictions and beliefs are religious in nature. Thus, I want to reflect on the Christian theological tradition and indicate how *moral* judgments that Christians attain on issues of human gene transfer are, or ought to be, informed and shaped by and partially dependent on specifically *theological* beliefs. In other words, I want to suggest that the moral decisions that Christians come to concerning whether or not to support the alteration of our genetic code depend partially on a religious context of meaning. This religious context can inform and authorize certain moral judgments that believers might make. By stating the matter this way, I do not mean to imply that one can separate moral and religious experiences. I intend only to claim that the two are distinguishable and then to indicate how one can influence or qualify the other. Thus, in the case at hand, theological convictions can provide perspectives on and engender attitudes about genetic manipulation. Furthermore, this religious context does not by itself determine moral decision making for Christians; there are, of course, a number of other background issues that function as presuppositions to moral judgments on human gene transfer. The following list is merely a sample of such issues: the goals and limits of medicine,[4] the meaning of suffering and illness,[5] attitudes about genetic disabilities,[6] and the relation between science and theology.[7] Generally speaking, if one or other of the genetic technologies to be discussed were found to be inappropriate on *theological* grounds because, for example, their use would usurp God's rights over creation, the presumption might be that these same technologies as a consequence would be judged *morally* unjustified. In fact, the claim would be made in this case that such interventions are arrogant attempts at "playing God."[8] On the other hand, if it can be shown theologically that certain kinds of human gene transfer are not contrary to the divine's final purposes for humanity, then it might be possible, *along with other evidence*, to judge these gene technologies as morally defensible. In the conclusion to my presentation I will state where I would stand morally on the various forms of human gene transfer.

SIX CENTRAL THEMES FROM THE ROMAN CATHOLIC TRADITION

Before pursuing the theological concerns, though, I will summarize briefly what I believe are the central themes that inform this issue from the official Roman Catholic perspective, i.e., from the magisterial teachings of recent popes, bishops, and the Second Vatican Council. In general, I find most official statements since Vatican II (1965) to be quite hopeful and favorable toward genetic science with respect to the issue of manipulating the human genome as long as certain moral boundaries are respected. What are some of these boundaries or themes that pertain to the specific issue of human genetic manipulation? There are six.

First, we are permitted to pursue various genetic manipulations as long as we respect the natural law, i.e., the moral law that is inscribed in the nature of humans and their moral acts. In the Catholic tradition the order of nature grounds human morality, and this morality is not only objective but also in principle capable of being known by all people of goodwill. As Cardinal Karol Wojtyla (Pope John Paul II) claimed in his book *Love and Responsibility*, a rational acceptance of the order of nature is at the same time a recognition of the rights of the Creator.[9] Concretely, the natural law requires that we respect the dignity of each human being, and thus the natural law would prohibit treating humans and embryos from the moment of conception as a means to some other end. Second, the official teachings from the Roman Catholic church express a strong ethic of stewardship. This ethic points to two things: a) we have a God-given responsibility for and toward all creation, including our bodies; and b) we are *not* the owners of our own bodies but only stewards over them, so we are *not* free to manipulate our genetic heritage (or nature) at will. Third, the human body is not independent of the spirit. Concretely, this means that we cannot expect to alter our genes without also altering the body's relation to our spiritual natures, i.e., who we are as a body-soul composite.[10] Fourth, genetic experimentation on human subjects, including embryos from the moment of conception, are permissible as long as "it tends to real promotion of the personal well-being of humans, without harming human integrity or worsening human life."[11] The informed consent from the one experimented on or from a legitimate surrogate is absolutely required for such experimentations. Fifth, there is a fundamental relationship between scientific research and the common good of society. This clearly indicates that all such efforts to manipulate the human genome involve not only ethical but also public policy implications. Finally, not every scientific advance necessarily constitutes a real human progress. Though John Paul II is not absolutely opposed to all forms of nontherapeutic genetic interventions,[12] there are some statements from the U.S. bishops' NCCB Committee on Science and Human Values that seem to limit genetic testing and genetic manipulation to instances in which there is an effective therapy or cure of a genetic abnormality for a patient or embryo.[13] Thus, genetic manipulation to influence inheritance that is not therapeutic but aimed at producing human beings selected according to sex or other predetermined qualities (eugenics or enhancement manipulation) is judged contrary to the personal dignity of the person and consequently contrary to the natural law. This last point in particular leads us to a consideration of the various types of what is called human gene transfer.

TYPES OF HUMAN GENE TRANSFER

Imagine a day when patients with defective genes that cause them great disability can walk into a clinic and be given an injection of engineered cells that contain the proper sequencing of the genes to cure their diseases. Or, imagine a day when prospective parents can simply walk into a clinic for assisted reproductive technology and preselect the enhanced genetic traits that their future child will have. This is not just science fiction; it will likely become reality in the not-too-distant future. Why? Medical scientists could conceivably develop four different types of human gene transfer from the results produced in the Human Genome Project. In other words, medical science will shortly have the capacity to alter our genetic code in four ways. The first two types are therapeutic in nature because their intent is to correct or prevent some genetic defect that causes disease. The other two types are not therapies at all, and many question whether they are part of medicine's goals as well. Rather, they are concerned with improving either various genetic traits of the patient him/herself (somatic cells) or with permanently enhancing or engineering the genetic endowment of the patient's children (germ-line cells).

The first kind of human gene transfer is somatic cell therapy in which a genetic defect in a body cell of a patient could be corrected by using various enzymes (restriction enzymes and ligase) to splice out the defect and to splice in a healthy gene. Medical scientists have already used a variation of this technique to help children who suffer from severe combined immune deficiency (ADA) by modifying bone-marrow cells, and a similar procedure was used in August 1999 for children who have Crygler Najjar syndrome, a genetic disease that causes fatal brain damage. Estimates are that between two to five thousand different genetic diseases are controlled by one gene, and these diseases afflict approximately 2 percent of all live births. Second, there is germ-line gene transfer therapy in which either a genetic defect in the reproductive cells—egg or sperm cells—of a patient would be repaired or a genetic defect in a fertilized ovum would be corrected *in vitro* before it is transferred to its mother's womb. In either case, the patient's future children would be free of the defect by permanently altering their genetic code.

Next there are the two kinds of nontherapeutic human gene transfer. The first kind is somatic enhancement engineering. In this type, a particular gene could be inserted to improve a normal trait, for example, the insertion of a new gene or an improved one to enhance memory. Second, there is germ-line genetic engineering in which existing genes would be altered or new ones inserted into either germ cells or into a fertilized ovum such that these genes would then be permanently passed on to improve or to enhance traits of the patient's offspring. In this last form of human gene transfer parents could design their children according to their own desires.

THEOLOGICAL THEMES

Whether or not Christian believers and Christian churches will support *morally* any or all of these types of human gene transfer will partially depend on where they stand on certain theological beliefs about God and humanity. Christians, at least, have regularly made sense out of their experiences of God and then communicated these interpretations to others by reference to a story that has been reformulated into certain doctrinal themes. Traditionally, these themes have been expressed in the Christian tradition in terms of creation, fall, incarnation, redemption, and eschatology.[14] The Christian story tells us that God is the creator of all that is and that we are from God and for God. However, sin or alienation from God, self, and others has entered the world due to human misuse of freedom. Yet, God has decided to bind divine history irrevocably to human history by becoming incarnate in human form (Jesus is fully human and fully divine). Through Jesus's preaching of the Gospel and his passion, death, and resurrection we are redeemed (redemption) and called to a new future in God's eternal kingdom (eschatology). Though all of these themes are important for our purposes, I will discuss only three of them, viz., creation, incarnation, and redemption. Again, my purpose in referring to these religious themes is to show how *moral* judgments on human gene transfer rely on and can be authorized by certain *theological* beliefs and interpretations.

Creation and Divine Providence

The doctrine of creation is actually a complex set of interpretations of who God is and how the divine directs human history and acts within it (divine providence). These theological interpretations have anthropological counterparts that attempt to understand both how we as created beings stand in the image of God (*imago dei*) and how we are to evaluate the significance of physical nature and our bodily existence.

Two different theological models of God, creation, and divine providence have been historically used in the great Christian tradition. Currently, Christians have used both models as a theological context in arguing morally for or against human gene transfer.

In one perspective God is viewed as the creator of both the material universe and humanity and the one who has placed universal, fixed laws into the very fabric of creation. This view of creation assumes that God's purposes for humanity, which are forever unchangeable, can be known by reflecting on the universal laws governing nature and humanity. As sovereign ruler over the created order, God directs the future through divine providence. As Lord of life and death, God possesses certain rights over creation, which in some cases have not been delegated to humans for their exercise.[15] When humans take it upon themselves to exercise God's rights, for example, those divine rights to decide the future or to change the universal laws that govern biological nature, they usurp divine authority and thus act contrary to God's purposes in creation and "play God."

If one adopted the theological positions held in this model, one would likely judge as human arrogance any attempt to alter the genetic structure of the human genome, possibly even for the therapeutic purpose of curing a serious disease. This assessment is confirmed in a *Time*/CNN poll (January 1994) on people's reaction to genetic research. Not only were many respondents ambivalent about genetic research but a substantial majority of the respondents (58 percent) thought that altering human genes *in any way* was against the will of God.[16]

In a second theological model God is not interpreted as the one who has created both physical nature and humanity in their complete and final forms. Rather, the divine continues to create in history (*creatio continua*). Consequently, God is not understood as having placed universal, fixed laws into the fabric of creation, and so the divine purposes are not as readily discernible as in the first model. God's actions both in creation and in history continue to influence the world process, which is open to new possibilities and even spontaneity.[17] Though there is some stable order in the universe, creation is not finished, and history remains indeterminate. Because creation was not made perfect from the beginning, one can discern certain elements in the created order, like genetic diseases, that are disordered. Because these disordered aspects of creation cause great human suffering, they are judged to be contrary to God's final purposes and so can be corrected by human intervention. Thus, therapeutic types of human gene transfer could conceivably be justified in this interpretation, though it might be difficult to justify morally the two enhancement types (somatic and germ-line).

As an anthropological counterpart to their interpretations of the divine, Christians have consistently understood all humanity to be created in the image and likeness of God (Gen. 1:26-27). However, the great Christian tradition has used at least two different interpretations of how humans stand in that image, and these diverse models almost inevitably lead to different moral evaluations about interventions into the human genome.

The first interpretation defines humanity as a steward over creation. Our moral responsibility, then, is primarily to protect and to conserve what the divine has created and ordered. Stewardship is exercised by carefully respecting the limits placed by God in the orders of biological nature and society.[18] It is easy to see how this model is consistent with the understanding of God as the creator who has placed universal, fixed laws into the very fabric of creation. If we are stewards over both creation and our own genetic heritage, then our moral responsibilities do not include the alteration of what the divine has created and ordered through nature. Our principal moral duties are to remain faithful to God's original creative will and to respect the laws that are both inherent in creation and function as limits to human intervention. In this scenario, most, if not all, forms of human gene transfer would be morally prohibited, though some room might be permitted for somatic cell therapy.

The second interpretation of the *imago dei* defines humans as created cocreators or participants[19] with God in the continual unfolding of the processes and patterns of creation. As created cocreators, i.e., as beings who do not create *ex nihilo* as God does,[20] we are both utterly dependent on God for our very existence and simultaneously responsible for creating the course of human history. Though we are not God's equals in the act of creating, we do play a significant role in bringing creation and history to their completion.[21] Proponents of this interpretation would almost certainly support somatic, and possibly even germ-line, gene transfer aimed at therapeutic ends, though it is highly questionable whether they would also justify attempts at enhancement gene transfer for somatic or germ-line cells.

A Christian interpretation of the significance and value of both physical nature and human bodily existence also plays an important role in arriving at moral judgments on genetic interventions. There are several different models of material nature that can shape one's moral position on human gene transfer, and Christian authors have made use of all of them. Each model attempts not only to interpret the nature of all material reality but also to understand the extent to which we can use human freedom to change our genetic heritage.

Daniel Callahan has argued that one of the most influential models of nature that operates in contemporary society is the power-plasticity model. In this view, material nature possesses little or no inherent value, and it is viewed as independent of and even alien to humans and their purposes. All material reality is simply plastic to be used, dominated, and ultimately shaped by human freedom.[22] Thus, the fundamental purpose of the entire physical universe, including human biological nature, is to serve human purposes. What is truly human and valuable are self-mastery, self-development, and self-expression through the exercise of freedom. The body is subordinated to the spiritual aspect of humanity, and humans view themselves as possessing an unrestricted right to dominate and shape not only the body but also its future genetic heritage. This view would be strongly inclined to justify morally almost any intervention into the human genome, regardless of whether its intent is therapeutic or enhancement.

The view of nature at the opposite extreme is the sacral-symbiotic model in which all material nature, including our genes, is viewed as created by God and thus considered as sacred. As created and originally ordered by God, human biological nature is static and normative in this understanding, and the laws inherent in it must be respected. Humans are not masters over nature but stewards who must live in harmony and balance with our material nature. Biological nature remains our teacher that shows us how to live within the boundaries established by God at creation. Since physical nature is considered sacrosanct and inviolate, any alteration of the human genetic code, even to cure or prevent a serious genetic disease, would probably be morally prohibited.

The final model construes material nature as evolving. Whereas there is some stability to nature and there are some laws that do govern material reality, neither this stability nor these laws are considered absolutely normative in moral judgments. Change and development are considered more normative than other aspects of nature, and history is seen as linear rather than cyclic or episodic.[23] The relation between material nature and human freedom appears as a dialogue that dynamically evolves over time. It is within this dialogue that humans learn how to use responsibly material reality as the medium of their own creative self-expression.[24] This model would seem to grant to humans the freedom and responsibility to intervene into our evolving biological nature to correct serious diseases even at the germ-line level. The reason is because such human efforts would not necessarily be judged as usurping God's final prerogatives or purposes in creation.

Incarnation

The fact that God took on human bodily form in the person of Jesus Christ has several implications for the discussion of genetic medicine. First, this doctrine serves as a context both for assessing the relation between body and spirit and for evaluating the significance of the body in moral decision making. In turn, these considerations have an impact on the question of what we judge to be uniquely or normatively human in moral analysis. Both issues function as presuppositions to moral judgments about the permissibility of human gene transfer.

If one separates, or even grossly distinguishes, body and spirit, there is the tendency to view our spiritual part as more normatively important or even as the solely unique characteristic of the human person. In addition, such a view will tend to hold that permanent alterations of the body, which would occur through the various types of human gene transfer, do not and cannot actually change the fundamental nature of humans. Dr. W. French Anderson, arguably the most influential human gene therapist in the United States, once remarked that he had been worried for years that we might end up altering our very humanness by methods of human gene transfer, especially those aimed at the germ line. However, he has recently decided that Plato was correct to view the soul and the body as two distinct entities.[25] By adopting this Platonic framework Anderson now believes that we cannot alter our fundamental humanness because, as much as we might permanently change our biological genetic code through gene transfer, we cannot change that which is uniquely or normatively human about us, viz., our soul or that which is beyond our "physical hardware."[26] Thus, ostensibly Anderson would justify morally both somatic and germ-line therapeutic interventions to alter permanently the human genome.

An opposing view is the position that holds that there is an intimate relation between body and soul. In this construal humans are viewed as embodied spirits or ensouled bodies.[27] Such an interpretation, then, would be far more cautious than the first about making a claim that we cannot permanently alter the nature of humanity through human gene transfer. The relation of body and spirit is one, but not the only, element of what makes up our fundamental human nature. Thus to alter radically this relation of body and spirit would imply the possibility of changing our nature in this view. Though proponents of this interpretation could support morally gene transfers aimed at the prevention or curing of disease, they would neither encourage nor support enhancement techniques. The reason would probably be that in the latter cases of enhancement interventions the chances of altering the body-spirit relation might be greatly increased.

Redemption

Christians believe not only that we are created, though fallen, beings but that we are also redeemed by God through the suffering, death, and resurrection of Jesus Christ. Thus, besides God's *creative* purposes or ends, the divine also has *redeeming* purposes for all creation, i.e., to bring all creation fully into God's kingdom. Christians have sometimes grossly separated the creative and redeeming purposes of God. One way to understand the relation between these divine activities has been to interpret redemption as not only a continuation of creation but the means by which creation itself is brought to completion by God.[28]

This framework raises the question of whether the technologies to alter our genetic code can ever be viewed as potential participations in God's redeeming actions toward humanity. Since Christians have interpreted humankind as created in the divine image, it has been possible to view genetic interventions as possible acts of *co-creation* with the divine. However, now the question is whether it is also possible theologically to view our technological activities as potential participations in or mediations of God's *redemptive* purposes. To answer this question requires a brief discussion of various theological evaluations of technology in general.

There are several evaluations of modern technology that could serve as the context for our moral judgments on therapeutic techniques to cure serious genetic diseases. First, there is the rather pessimistic view of technology, an example of which was the position taken by the late Jacques Ellul.[29] Its characteristics include a very skeptical attitude toward any real benefits from technology and a great sensitivity to the potential evils that will come from its development and use. Technology is viewed as a threat, impersonal, manipulative, and alienating, and thus it does not and cannot possess the inherent potential to share in the divine purposes of redemption, which are viewed as personal, salvific, and holistic. In the end, this view would probably not support morally any attempt to alter our genetic heritage.

The opposite extreme is an overly optimistic view of technology and its potential achievements. Its hallmarks are a focus on the liberating function of technology through progress and human fulfillment and an emphasis on greater freedom and creative expression. Some, like the Jesuit paleontologist Pierre Teilhard de Chardin,[30] have closely linked technology and spiritual development and thus have viewed technology as clearly possessing the potential to cooperate with God's work. Some who have adopted this position have been quite supportive morally of most forms of human gene transfer, including those forms whose primary purpose is to enhance or engineer our genes.

The final position seeks to steer a middle course between the two extremes of pessimism and optimism. Similar to the first view, proponents are cautious about and critical of many features of modern technology. However, like the second view these proponents also offer hope that technology has the potential to be used for humane moral ends, but technology must be redirected in its uses for these ends to be realized. There are two forms of this moderate position currently held by theologians that I would like to analyze quickly. Among other things, these views are distinguished by how they connect causally sin with disease and death. In other words, these positions differ depending on how one interprets St. Paul's passage in the Letter to the Romans (5:12): "It is just like the way in which through one man sin came into the world, *and death followed sin*, and so death spread to all men, because all men sinned."

The first position links causally the introduction of both death and all disease, including genetic disease, to the entrance of sin into the world. The role of medicine, then, is to intervene to overcome these effects of sin, and these medical interventions, including those aimed at genetic therapy, are construed as mediations of God's redemptive activity. In this same view, however, all forms of human gene transfer whose primary purpose is to enhance or engineer the human would be at least morally problematic on theological grounds. Why? Because these interventions would not alleviate any condition that can be linked causally to the entrance of sin into the world. Their purpose would be to enhance the patient or his or her progeny, not to overcome the effects of the Fall.[31]

The second form of the moderate position does not link causally sin with disease and thus does not identify disease as such as one of the effects of the Fall. Rather, it understands diseases (and for that matter, death) as the natural result of being part of the material world where decay and entropy are facts of the created world, though sin may very well adversely affect our experiences of these realities.[32] That does not mean that God wills or permits these ill effects *as part of the final divine ordering* of the universe; in fact, they are judged to be contrary to God's ultimate purposes. The Protestant theologian Ronald Cole-Turner has adopted a position similar to this one.[33] He has argued that modern technological developments in genetics can have the potential for participating in God's redemptive activities. He has reasoned that, when this technology is *aimed at preventing or curing serious genetic diseases* that are deemed contrary to God's final purposes for humanity because they cause great human suffering, this technology can participate in God's redemptive purposes by making whole and healthy what was once disordered and destructive. Cole-Turner, like the first form of this position, however, does not seem to support morally human gene transfer whose primary purpose is enhancement, not therapy.

CONCLUSION

Given the specifically theological context discussed, it is now time for me to state where I would stand morally on the various types of human gene transfer that will no doubt result from the HGP. It is my judgment that significant scientific and technical difficulties remain to be solved with most forms of human gene transfer. For example, the fatal experiment in September 1999 on Jesse Gelsinger, the eighteen-year-old who had been injected with engineered genes by the University of Pennsylvania researchers to cure the boy's rare liver disease, and the recent National Institutes of Health order to toughen the rules on the reporting of deleterious side effects of gene therapy[34] only indicate again that we may well be years away from when many of these therapies can be shown to be both safe and effective. In addition, there are a number of public policy problems with these interventions as well. Consequently, *at the present time*, I am opposed morally to all types of gene transfer, at either the somatic or germ-line level, whose only purpose is to enhance or engineer human traits. On the other hand, I can morally support therapeutic gene transfers on somatic cells, when and as long as these scientific techniques can be shown to be safe and effective. In addition, I would also argue on Christian theological grounds that once the scientific, public policy, and moral difficulties can be resolved with germ-line therapeutic interventions, we may cautiously move forward with this type of genetic therapy as well, i.e., as long as they can be proven to be both safe and effective. In other words, based on my theological interpretations of the Christian doctrinal themes and their anthropological counterparts already discussed, I conclude that in principle the two types of human gene transfer aimed at therapy or prevention, viz., somatic and germ-line interventions, are not fundamentally contrary to God's purposes for humanity. To use them is not necessarily to arrogate to ourselves various functions and tasks that properly belong only to the divine. Rather, they have the potential or capacity to mediate God's final purposes for humanity. Consequently, their use for the moral ends of preventing or curing serious genetic diseases can be a means of properly exercising human responsibility.[35]

NOTES

1. Aaron Zitner, "Humans Need Fewer Genes Than Thought to Survive," *Los Angeles Times*, February 11, 2001, A1 and A44.

2. "Genome Project Has Sequenced Two-Thirds of Human DNA," *Los Angeles Times*, March 30, 2000, B2.

3. Thomas F. Lee, *The Human Genome Project: Cracking the Genetic Code of Life* (New York: Plenum, 1991), 295.

4. Eric Juengst, "The NIH 'Points to Consider' and the Limits of Human Gene Therapy," *Human Gene Therapy* 1 (1990): 426 and 431.

5. Lucien Richard, O.M.I., *What Are They Saying About Genetic Engineering?* (New York: Paulist, 1992).

6. Ian G. Barbour, *Ethics in an Age of Technology* (San Francisco: Harper-Collins, 1993), 196.

7. Zachary Hayes, O.F.M., *What Are They Saying about Creation?* (New York: Paulist, 1980), 7–20; James M. Gustafson, *Intersections: Science, Theology, and Ethics* (Cleveland, Ohio: The Pilgrim Press, 1996); and Robert J. Nelson, *Human Life: A Biblical Perspective for Bioethics* (Philadelphia: Fortress, 1984), 167–70.

8. For an important article on the various meanings of "playing God," see Allen Verhey, "'Playing God' and Invoking a Perspective," *The Journal of Medicine and Philosophy* 20 (1995): 347–64.

9. Karol Wojtyla (John Paul II), *Love and Responsibility*, trans. H. T. Willetts (New York: Farrar, Straus and Giroux, 1981), 246.

10. John Paul II, "Biological Research and Human Dignity," *Origins* 12 (1982): 342.

11. John Paul II, "The Ethics of Genetic Manipulation," *Origins* 13 (1983): 388.

12. Ibid., 388–89.

13. National Conference of Catholic Bishops/Science and Human Values Committee, "Critical Decisions: Genetic Testing and Its Implications," *Origins* 25 (May 2, 1996): 769, 771–72.

14. Elsewhere I have argued that these five themes comprise the horizon of Christian religious intentionality and constitute what is unique or specific about Christian ethics. See James J. Walter, "Christian Ethics: Distinctive and Specific," *The American Ecclesiastical Review* 169 (1975): 483–84.

15. See Josef Fuchs, S.J., *Christian Morality: The Word Becomes Flesh* (Washington, D.C.: Georgetown University Press, 1987), 39–61; and Jan Jans, "God or Man? Normative Theology in the Instruction *Donum vitae*," *Louvain Studies* 17 (1992): 56–63.

16. Philip Elmer-Dewit, "The Genetic Revolution," *Time*, January 17, 1994: 48.

17. Ian G. Barbour, *Issues in Science and Religion* (New York: Harper and Row, 1966), 449.

18. Thomas A. Shannon, *What Are They Saying about Genetic Engineering?* (New York: Paulist, 1985), 21.

19. James Gustafson has preferred to describe our role in creation as *participants* rather than as *cocreators*. He argues that the divine continues to order creation, and we can gain some insight into God's purposes by discovering these ordering processes in nature. See his *Ethics from a Theocentric Perspective, Vol. II* (Chicago: University of Chicago Press, 1984), 294.

20. Philip Hefner, "The Evolution of the Created Co-creator," in *Cosmos as Creation: Theology and Science in Consonance*, edited by Ted Peters (Nashville: Abingdon, 1989).

21. Ted Peters, "'Playing God' and Germline Intervention," *The Journal of Medicine and Philosophy* 20 (1995): 377–79; and Ann Lammers and Ted Peters, "Genethics: Implications of the Human Genome Project," in *Moral Issues and Christian Response*, edited by Paul T. Jersild and Dale A. Johnson (New York: Harcourt, Brace and Jovanovich College Publishers, 1993), 302.

22. Daniel Callahan, "Living with the New Biology," *Center Magazine* 5 (1972): 4–12.

23. Shannon, *What Are They Saying*, 37.

24. Norris Clarke, S.J., "Technology and Man: A Christian Vision," in *Science and Religion: New Perspectives on the Dialogue*, edited by Ian G. Barbour (New York: Harper and Row, 1968), 287–88.

25. W. French Anderson, "Genetic Engineering and Our Humanness," *Human Gene Therapy* 5 (1994): 758.

26. Ibid., 759. The reverse type of reductionism of the normatively human, of course, is the one adopted by a research biologist at UC San Diego. When the drosophila fly's entire genome was recently mapped and sequenced and then its genes likened to the Rosetta stone for the mapping and sequencing of the human genome, Dr. Charles S. Zukor stated, "We are nothing but a big fly." See Robert Lee Hotz, "Full Sequence of Fly's Genes Deciphered," *Los Angeles Times*, March 24, 2000.

27. See John Paul II, "Biological Research," 342; idem, "The Ethics of Genetic Manipulation," 388; and Johnstone, "La tecnología genética," 307–08.

28. For example, see Karl Rahner, S.J., "The Order of Creation and the Order of Redemption," in *A Rahner Reader*, edited by Gerald A. McCool (New York: Seabury Press, 1975), 190–96.

29. Jacques Ellul, *The Technological Society* (New York: Knopf, 1964).

30. Pierre Teilhard de Chardin, *The Phenomenon of Man* (New York: Harper and Row, 1959).

31. See Scott B. Rae and Paul M. Cox, *Bioethics: A Christian Approach in a Pluralistic Age* (Grand Rapids, Mich.: Wm. B. Eerdmans, 1999), 118–27.

32. This position is more informed by St. John's Gospel than St. Paul's Epistle. In John 9:1–3, the evangelist writes, "As he went along, he saw a man who had been blind from birth. His disciples asked him, 'Rabbi, who sinned, this man or his parents, for him to have been born blind?' 'Neither he nor his parents sinned,' Jesus answered, 'he was born blind so that the works of God might be displayed in him.'" I would like to thank my colleague, Jeffrey S. Siker, for pointing out this biblical text to me.

33. Ronald Cole-Turner, *The New Genesis: Theology and the Genetic Revolution* (Louisville, Ky.: Westminster/John Knox Press, 1993), 80–97.

34. Marlene Cimons, "NIH to Order New Reports on Past Gene Therapy Cases," *Los Angeles Times*, February 24, 2000, 1, 14.

35. An earlier version of this article was published by the *New Theology Review*.

7

Human Cloning: Religious and Ethical Issues

Thomas A. Shannon

Undoubtedly, the mapping of the human genome will be a boon to science, medicine, and anthropology, among other disciplines. This map will provide a clearer lens with which to examine the question of what *being human* means, will direct us where to look for anomalies that cause disease, and will greatly assist in correcting those errors. We are genuinely on the edge of a new revolution in medicine, one that will provide access to the very structure of our nature. We can literally reach inside ourselves, remove genes, and either correct or replace them. Such power, though truly awe-inspiring, is also truly frightening.

Yet these achievements bring risk as well. Some fear that the new genetics will inspire a new eugenics and that the Human Genome Project will set a genetic standard by which all humans are measured and evaluated. Again, the individual seems to be in danger of being subordinated to the type. Additionally, new developments in behavioral genetics are building up suggestive evidence for the role genes play in all manners of human behavior, from sexual preference to choices of political perspectives and marriage partners. In learning more of what it means to be human, will we become less human in the process?

The embryo-division experiment by Dr. Jerry N. Hall and Dr. Robert Stillman[1] of Washington University and the brief, but lively, discussion that followed it once again focused on many of the thematic issues raised by ethical, scientific, religious, and cultural debates over genetic engineering: human power over nature, intellectual arrogance, the technological imperative, action before thought, the degradation of human beings, and the violation of their unique genetic structure. Yet, the experiment also offered the promise of benefits: advanced knowledge of the developmental process of preimplantation embryos and the further development of new responses to infertility. Even though the university research ethics committee cleared this experiment, Hall and Stillman have retired from the embryo-division business, at least for the present. Public and professional reaction seemed to be quite strong against such embryo division.

The most recent chapter of the genetics debate was written by Dr. Ian Wilmut of Scotland when, on February 22, 1997, he announced that he had successfully cloned a lamb which he named Dolly. This was the first successful application of cloning, or nuclear-transfer technology, in a mammal. More significantly, the nucleus of the cell that produced Dolly came from a six-year-old ewe, thus showing that scientists may *turn on* genetic instructions that were previously thought to be irreversibly turned off.[2]

Every cell of one's body contains all the genetic information needed to make a whole other being. Very early in embryonic development, however, this information is selectively turned off, and various cells become committed to becoming specific body parts through a process called restriction. The technological breakthrough of Dolly is that Dr. Wilmut succeeded, after 277 attempts, to have the DNA from a six-year-old cell turn on and be the source of the genetic information that eventually led to the development of Dolly. After the nuclear transfer was completed, an electrical charge was applied to the fused cell so that its contents would emerge from the nucleus and the process of cell division would begin. Two elements of this process are critical: (1) the DNA that had been turned off was turned on again, and (2) the cell was six years old, which showed that the restriction process could be reversed even in adult cells.

This experiment has set off yet another massive international debate on bioethical issues. In this chapter, I will review several religious and ethical perspectives in the cloning debate and conclude by arguing for an extremely narrowly drawn case favoring some forms of cloning, while rejecting others.

THE CULTURAL PRESENTATION OF CLONING

In this section, I wish to present some elements of the cultural context in which cloning was presented. This section will also introduce elements of the cloning debate that later sections will develop.

Presentations in the Media

Because genetics, whether medical, agricultural, or animal, is big business, many stories on Dolly's unique origin were reported in the economics section of newspapers.[3] That is not unusual now, for news of genetic developments have regularly been reported in that section of newspapers ever since gene companies went public over a decade ago. Some of this reporting vacillated between praise for the technology and concern about whether ethical criticisms might depress the growing biotechnical market. While on the one hand biotech leaders affirmed the immorality of cloning a full human, they were also concerned that such ethical tut-tutting not go too far lest this rapidly growing industry be harmed, particularly in its agricultural and animal applications where cloning techniques are routinely used. Thus, Dolly has raised the science-business ethics debate again, but this time major financial interests of the biotechnical-industrial complex, new players in such debates, are mediating the debate.

Another interesting dimension appearing in the science section of newspapers was the reported comments of some senior American scientists who dismissed the significance of the research by suggesting that one could not write a grant proposal to clone a sheep because the scientific question was unclear. Another suggested that the project was mundane and merely technological.[4] Is this professional jealousy? Remember that the United Kingdom is now two up on American science: Patrick Steptoe and Robert Edwards had the first *in vitro* fertilization (IVF) baby, and now Wilmut has the first cloned mammal. While it is true that scientists are motivated by the quest for knowledge, they are also motivated by the quests for patents and the financial rewards that come from them. So perhaps it was no accident that a few days after the announcement of Dolly's creation, American scientists announced that they had cloned two rhesus monkeys, though they used embryonic cells, not the more difficult older adult cells that were Dolly's progenitor. The cloning sweepstakes are wide open.

A final note about cloning appeared in the entertainment section which discussed various cloning movies.[5] Of note was the split between comedy and science fiction terror stories. While *Multiplicity* did not do well at the box office, it did show some practical applications of the technology to help resolve domestic complications. It also observed, though, that sometimes when one makes a copy, the second one is not quite as sharp as the first. *Twins* showed the comic possibilities by presenting identical twins who were not quite identical. This was blamed on making one twin from leftover DNA, what is technically referred to as *junk DNA*, the sections of DNA whose function we do not know. *Jurassic Park* told the familiar morality tale of the evils of commercialization of technology. The movie ended with everyone but the "mad" scientist living happily every after, but much material remained to clone a sequel, which of course is now available. The more recent offering *Gattaca* explores difficulties of an individual who does not meet the genetic standards of the "perfect society." This movie was preceded by an advertisement campaign that resembled advertisements for genetic clinics that would provide a baby with the characteristics chosen by the parents. Only when one looked at the very fine print at the bottom of the full-page advertisements, did one realize that this was science fiction, not science.

Presentations of the Cloning Debate

What I term *ethics hysteria* has dominated much media coverage. This ethics hysteria takes the worst possible ethical and most technically improbable scenario and builds the case for rejecting cloning on that basis. Probably the best single example of such a hysterical presentation of cloning was the February 10, 1997, cover of *Der Spiegel,* the German equivalent of *Time.* Marching down the cover were multiple copies of Adolf Hitler, Albert Einstein, and Claudia Schiffer. While not all of the figures may represent everyone's worst case scenarios, the technology of cloning is presented as replicating an infinite series of beings who are not only genetically identical, but more importantly are multiple copies of the very same person.

This scenario raises another dimension of the cultural presentation of the cloning debate: genetic reductionism. This position argues what the cover of *Der Spiegel* presented graphically: by simply replicating my genetic code, I am thereby replicated. However, cloning does create an offspring that is genetically identical to the donor of the DNA. But what follows from that? What follows is genetic identity: the clone is genetically identical to its source. It may even look identical. The hidden premise of genetic reductionism is that all that I need to make me *me* is my genetic profile. Such an argument ignores the fact that the clones are two distinct individuals and have their own distinct environment in which they are raised, to say nothing of ignoring any transcendent or personal dimension of the human. Thus, the clone and the donor quickly begin to part ways.

To understand this better, consider human identical twins, who are in fact clones of nature. The fertilized egg divides, resulting in two distinct individuals with identical genetic profiles. We know that many studies on identical twins have shown that they share many interests and similarities, even when raised apart in radically different environments. Perhaps these studies are what drive the fantasy that creating clones of Michael Jackson or Michael Jordan would create beings with exactly the same abilities and interests present, perhaps even at the virginal conception.[6]

Although our genetic heritage strongly influences us, environment shapes our lives too. Suppose Michael Jordan's clone were raised in an environment or culture in which the main and perhaps preferred career option for African

Americans was not professional sports. Would a genetic determinism be at work that would impel him to play basketball no matter what? This seems to be the great flaw in the cloning debate: genetics will win out no matter what. This is simply not true. Nonetheless, genetic determinism is paradoxically both the assumption of the outcome of cloning, as well as the major argument against human cloning.

What this brief survey reveals is that various agendas are at work in the cloning debate: economic, cultural, scientific, and political. We have no canonical presentation or analysis of cloning, nor can its discussion be cleanly extracted from a variety of cultural perspectives. But this overview directs us to look at and attend to certain issues as we seek to evaluate this stunning development in genetics.

RELIGIOUS PERSPECTIVES

I begin with a consideration of some religious perspectives because often religion is perceived as the great naysayer, the enemy of science, and the protector of the status quo, preferably the one of several centuries ago. While I disagree with such stereotypes, some religious themes do tend to be less supportive of interventions into creation or human life. In this section, I wish to discuss two religious themes—humans playing God and humans created in the image of God—and to show traditional usages of these terms, as well as other ways to interpret these themes that allow some leeway in their interpretation.

Two Traditional Religious Themes

Humans Playing God

We often use the term *playing God* as a way of arguing that humans have overstepped their boundaries. This term suggests that a clear demarcation exists between the roles of God and humans and that there are areas of life where God rules, where God is in charge, and where humans ought not enter. The term evokes an omnipotent God who is the Creator of all and who commands all. The term also evokes the image of God as "God of the Gaps," that is, the God who is invoked when all else fails, or when we have exceeded our limits, our knowledge is at an end, and our powers frustrated. Thus, it is most clearly in the gaps that God rules, and it is in the gaps that God's power is most clearly evoked. Here, God reigns supreme, and, here, we cannot play God.

Of course, as knowledge increases, the gaps grow smaller and smaller, and as a result, God's reign shrinks, God's power becomes lessened, and

God becomes less necessary. Then humans step into the recently vacated gaps and play God by exercising the powers in the gaps previously thought only God's. Cloning surely symbolizes such a disappearance of a gap and an exercise of new powers.

Such a vision of human intervention into nature is hardly Christian. It is certainly much more Greek, much more resonant with the myth of Prometheus, who in stealing fire from the gods and giving it to humans became like the gods and thus played God. However, he suffered the fate of one who usurped the power of the gods.[7] Were the God who is suggested by this version of playing God actually this fearful of sharing creation, assumedly God would never have created in the first place. Why spoil the way things are!

Perhaps a better rendering of playing God is to learn as much about God as one can and then to play God by acting as God acts.[8] Minimally, this might mean that we are to be creative as well as generous in our creativity and to keep covenant with our God and our creations. To affirm this is to surrender full control because we are not God. But it is also to assert a profound relation between ourselves and the rest of created reality. We play God by imitating God—no small task.

Two immediate consequences follow from this. First, this image of playing God does not set up a kind of competition between God and humans. The theme is stripped of its traditional mythological overtones and given a chance to return to a version much more faithful to the Judeo-Christian tradition. Second, in principle, such an understanding of playing God does not prohibit interventions in created reality. The moral element here would focus on that kind of intervention. A more helpful hermeneutics for understanding the term playing God might in fact be genuine play—and the nuance is that play cannot be purely instrumental, for then it is no longer play but work. And although the book of Genesis describes creation as a labor from which God rested[9] and the book of Job presents creation as a kind of civil engineering project,[10] the book of Wisdom describes creation as a form of play.[11]

Humans Created in the Image of God

A second religious theme is that of the human created in the image of God. A traditional understanding of this theme is that of humans as stewards who conserve and protect what God has created. Typically, one does this by respecting both the design of creation and the limits that God has placed on both the orders of biological nature and human society. Because this God designed the universe according to a plan and indeed embedded this plan into nature, the responsibility of a steward is to remain faithful to this plan and conserve it.

Such an interpretation of the image of God in human beings is a conservative one that, while not totally opposed to all interventions, is focused more on recognizing limits and maintaining boundaries. This is not done because of a lack of Promethian hubris, but rather out of a sense of genuine humility, a recognition of one's place before God and a sense of how one is to live out one's vocation in the world.

But another understanding of the image of God in man is one suggested by Philip Hefner: the created cocreator.[12] This phrase is important on two levels. First, it identifies humans as created. That is, because we are created, we are dependent on God for our present and continued existence, and we are not God's equals as creators. But, second, we are cocreators. We become participants with God in the continuous evolving of both nature and history. We have a responsibility both for the development of each and for our neighbors as we seek to further the divine work of creation. Such a view clearly allows a much expanded view of human intervention into the world. As evolving, the world is a work in progress, and its fulfillment is partially dependent on our interacting with it through the creative use of our freedom.

Variations on These Two Themes

I would argue that a better reading of both the terms *playing God* and *humans created in the image of God* is that of created cocreators who, enjoined with the task of playing God, do so by helping to bring creation to its *final* fulfillment. Such a vision of God and one's relation to God suggests, however, some cautionary notes. This vision of creation is not one of instrumentality or control, two values highly prized in technology and our culture. Should the context of discussions about cloning reveal a tendency to such values, I think one could mount a strong religious critique against cloning because of the problematic nature of such values in human life. Let me illustrate this through an analogous case.

Several years ago a man underwent a reversal of his vasectomy so he and his wife could conceive a child in the hope that the child might serve as a source of blood marrow for their older daughter who was dying of cancer. The reversal was successful, and he and his wife conceived a child. The infant's blood marrow was compatible, and at age eighteen months, blood marrow was transferred from her to her sister. The therapy was successful, and both are living very happy lives. Although the parents most strongly argued that they did not see their newborn as a source of parts and although this case clearly does not involve cloning, I cannot help but think that it is an example of creating one human being to serve another's needs, clearly one of the more prominent cloning scenarios. What bothers me about this case and my extrapolation of it to cloning is that the majority of citizens either agreed with the family's decisions or argued that it was the family's right to do as it pleased. This says to me that our society already has a mind-set or cultural disposition to accept one of the most frequently mentioned scenarios in cloning: replacement parts.[13]

Another dimension of this same problem was recently reported in the *New York Times.* The story reports that doctors at Columbia-Presbyterian Medical Center in Manhattan mixed sperm and eggs to make various embryos with different backgrounds: "The idea was to allow prospective parents to select embryos whose parents resemble them physically or have the same ethnic background and are well educated—the best possible sperm and egg donors for those who cannot have babies of their own." The cost is $2,750. Many of these embryos are made with leftover eggs from IVF or eggs obtained from donors when the planned recipient changes her mind. The eggs are fertilized with sperm from commercial banks selected for particular characteristics. Such a practice is certainly another direction in the mechanization of reproduction and the commodification of human body parts and human embryos. And selection of embryos is based on desired characteristics, which brings us another step closer to the objectification of humans through their being valued for their function or for their value in a market. Again, such a mind-set helps create a cloning mentality.

Many current arguments for human cloning revolve around some sort of use of human clones for the benefit of the cloner (for want of a better term). Thus, the scenarios we hear of are clones as sources for spare parts, specialized social or work functions—usually the menial ones we do not wish to do—and vanity reasons—one of a good thing is never enough. Common to all of these arguments is the reduction of humans to means only, a rejection of the dignity of both cloner and clone, and essentially a commodification of the human clones. Thus, in these discussions, a certain instrument mentality is at work. While perhaps born of desperation or extreme need, the mind-set is, nonetheless, instrumental. The concepts of playing God and acting as a created cocreator do not, in my judgment, sit well with this mentality. Here, we have control and use, not freedom and appreciation.

THE ETHICAL DEBATE OVER CLONING

In the debate over human cloning, a distinction must be made between full human cloning—research directed to produce an adult clone—and research on cloning up to, but not including, implantation in a uterus. The distinction is important because of two elements. First, research on the human preimplantation embryo could be important for learning about embryonic development or about the reproductive process. Second, different moral claims can be made about the preimplantation embryo and the individuated human embryo.

Cloning and the Preimplantation Embryo

One of the claims about cloning is that it violates individuality or the individual's right to a unique genetic identity. First, it is important to distinguish between genetic uniqueness and individuality. A preimplantation embryo is genetically unique in that it is a new combination of the genes from the mother and father. But it is more precise to say that this preimplantation embryo represents the next genetic generation precisely because it has not yet reached the developmental stage of restriction in which the cells become irreversibly committed to forming specific body parts in a particular body. Or, to say this more technically, there is as yet no differential gene expression. I argue that the preimplantation embryo presents itself as the biological equivalent of what the medieval philosopher Duns Scotus called, in philosophical terms, the common nature. Because the cells of the preimplantation embryo have the capacity of totipotency—the ability to become any part of the body—they are most properly designated as representing what is common to humanity.[14] The genetic structure they possess is generic to the species but is not yet identified with a particular individual, thus the term common nature. Though the cells of the preimplantation embryo possess a biological and teleological unity that will eventuate into a single human being, until these cells lose the capacity for totipotency through the process of restriction and become differentially expressed, we do not have what Norman Ford has called an "ontological individual."[15] Such individuality is biological in that the whole organism can no longer be divided into parts, each of which could become another organism as was the case before restriction. Such individuality is also philosophical in that this being is a single being with the potential to become a moral agent, an individual responsible for his or her own acts.

Given what we know about embryogenesis, a more precise way to describe this being at this stage is either the establishment of the next genetic generation or the establishment of the common nature. That is, while it is correct that the preimplantation embryo contains the appropriate genetic information for that organism's development, that genetic information is not necessarily associated with a specific individual and cannot, therefore, claim moral privilege through such an association. The genetic uniqueness is associated with what is common to all—human nature—not a particular individual because such an entity does not yet exist. The claim of the moral relevance of individuality is appropriately made of the preimplantation embryo only after the process of restriction has occurred, that is, after we have the only individual who in fact will emerge from the constriction of the common nature to this particular individual. This is the individual, all things being equal, who will become the agent of acts. Thus, while the fact that the preimplantation embryo manifests the human genome is morally relevant, it does not have the same moral relevance as individuality because the genetic status is associated with what is common to all, not what is unique to a person. The fact that the individual who emerges from this process may factually be the only one with this genotype also is of less moral relevance because we know, through Dolly, that such a genotype is in principle replicable. The focus of moral attention must be, in my judgment, on the individual manifested through this genotype, not the genotype itself.

What then are we to think of research on the preimplantation embryo either by dividing its cells or by nuclear transplantation? First, let us examine the dividing of the cells of the preimplantation embryo into separate entities. What one has with the preimplantation embryo is a teleologically united cluster of cells that has the capacity to become a distinct or ontological individual. The preimplantation embryo is neither all of humanity nor a particular human, but is the common nature out of which a particular, individual human can develop.

Therefore, to divide the four, eight, or sixteen cells of the preimplantation embryo into separate cells is not, in the memorably inaccurate phrasing of Germain Grisez, "splitting themselves in half."[16] Rather, it is to divide the whole organism into parts that themselves can become wholes. To do so is not to divide an ontological individual nor to violate that entity's distinct individuality. Although the preimplantation embryo is alive, carries the human genetic code, is genetically and physically distinct from its mother and father, and represents the next genetic generation, nonetheless such physical features of the preimplantation embryo do not constitute the philosophical basis for a claim of absolute value for this entity because there is as yet no subject or person on whose behalf to make the claim. In other words, no individual exists.

Second, let us consider cloning, that is, nuclear transplantation. Whole-organism cloning, as distinct from gene and cell cloning, takes the nucleus of an adult cell and puts it in the enucleated cell of another organism. The purpose is to genetically replicate the adult organism from which the nucleus came. The key to cloning is that, while each adult cell contains all the DNA necessary for the development of an entire organism, not all of that DNA is expressed. Such differential gene expression makes possible the development of the individual body parts and organs. The technical key to the success of cloning is to discover whether these unexpressed genes in the adult cell can be *turned on* and produce a genetically identical organism.

The common argument against cloning is that it violates the genetic uniqueness of the preimplantation embryo. This claim of moral standing based on genetic uniqueness in the preimplantation embryo cannot be sustained, for there is no subject of whom the claim can be made, as previously argued. Additionally, even if the cloning of humans were to succeed, what is replicated is the genetic structure, not the individual. No one claims that genetically identical twins violate each other's right to genetic individuality by virtue of bearing the same genetic structure. This is because, I argue, the more critical moral claim is that of individuality that is biologically secured only after restriction. Genetic uniqueness and its relation to identity is important for questions of lineage, but it is not the totality or even the basis of individuality.

How might one evaluate the morality of acts performed upon the preimplantation embryo whether obtained by its division or cloning? I suggest an examination of the object, the intention, and the circumstances of the act (particularly the circumstances of the end, the way the act is performed, the likely success of the act, and the circumstance of place).[17]

With respect to the object, i.e., the preimplantation embryos, these entities have a premoral value in that they are living, bear the human genome, and have a teleology directed to the moral category of personhood. However, because no individual subject exists of whom a claim can be made, no violation of individuality or personhood occurs. This premoral value must be judged in the light of other premoral and moral goods such as the benefits to come from research on these entities and the good of assisting in reproduction. I conclude that such goods—for example, experiments to discover the mechanisms to turn unexpressed genes on, and to study early embryonic development, or other studies to help enhance fertilization—could outweigh the claims of protection of the preimplantation embryo and that research, whether by division of the cells of the preimplantation embryo or cloning, could be done on the human preimplantation embryo before the process of restriction.

Considerations of the likely success of the act of cloning are difficult to calculate, for one genuinely does not know the full outcome or range of consequences that may follow the first experiment. This suggests that one very carefully consider the intentions as well as the purpose at which one aims. If the purpose of the cloning is to learn more of early cell development to aid in IVF, one could accept a lower level of success because the purpose is narrow and focused on internal development of the cells. Experiments attempting to turn on unexpressed genes could also be justified even though the lack of success may be low. The end of the experiment is focused on internal mechanisms of the gene. Further experiments that seek to apply such knowledge, however, would have to be very carefully examined in light of the end and intention.

I would interpret the circumstance of place to address the question of the priority of such research in relation to other needs in health care. Because such research is so expensive and applicable to only a narrow range of cases, a strong argument can be made against such research. If one broadens the argument to understanding the mechanisms of gene expression, then the range of application may be much broader—e.g., a better understanding of the immune system—and a different moral argument can be made. What is important is that the criterion of the circumstance of place makes us look to the setting of the research and its location in the full range of health care services as an appropriate source for moral evaluation of the research we wish to undertake. Such a criterion would, I think, assign a lower priority to cloning.

Thus, given an extremely careful consideration of both the moral status of the preimplantation embryo and the significance of the research involved, an argument can be made to justify such research up to the time when the process of restriction would occur, approximately two weeks after fertilization. Up to this time, the preimplantation embryo does not have the claim of true individuality and thus would lack the full moral protection associated with personhood, which cannot occur at least until such individuality is established. The moral argument justifying such research is, in my judgment, a very narrow one and one that must be examined on a case-by-case basis and surrounded by appropriate safeguards to prevent unwarranted extensions.

Individuality takes moral precedence over genetic uniqueness and is the key to the ethical analysis of research on the preimplantation embryo. Though I am factually the only one who bears my genetic profile, my genetic profile is not in principle unique to me. Genetic identity can be replicated either *in vivo*, through a natural cleavage of the preimplantation embryo into genetically identical twins, or *in vitro*, either through division of the cells of the preimplantation embryo or through organismic cloning (now done in a sheep and two rhesus monkeys). My genome is significant because it constitutes the establishment of my bodiliness, my human nature, and gives the basis for tracing my lineage. But more significant is my individuality both in the sense of indivisibleness and in the sense of the subject of moral acts. For it is only *I* who cannot be divided into parts, who can personalize that genetic structure, and who can transcend that genetic structure through personal acts. The absence of such individuality in the preimplantation embryo is a key element in the justification of the lack of its absolute protection as well as the possibility of some research, just as the presence of such individuality is a significant feature of its moral evaluation and is the basis of the prohibition of other types of research.

Cloning Scenarios

Replacements

One scenario commonly discussed is replacing a child who died or generating replacement parts through cloning one's identical twin. While this cannot technically be done at this stage, such discussions raise profound religious and philosophical issues of human dignity. What do such discussions say of the status of the clone? Has not this being simply been reduced to a means to an end? Has not this being's essential humanity been dismissed by reduction to the status of replacement? Such replacement talk avoids the fact that such entities are living beings with the human genotype while simultaneously reducing such beings to objects or commodities. The problem is that such talk assumes that *we* are not *they*, though both are genetically identical. But such an assumption is wrong precisely because it is the *we-ness* of our genes in the *they-ness* of their genes that we desire. How else could *they* be replacements for *us* if *they* are not genetically identical to *us*? To reduce such beings to commodities is to do the same to ourselves. Commonality of genetic identity suggests commonality of dignity as well as commonality of fate. Or to quote John Donne: "Never send to know for whom the bell tolls; it tolls for thee."[18]

This practice of cloning either does not stop to consider or simply rejects the humanity, individuality, and personhood of the clone. A clone of a human is a human and will be a person as I am—not the same person, but his or her own person. I do not know what others would call such an entity, but I would have to call it a person—at least after it has passed through some developmental markers such as the ones suggested in the preceding section. And the clone would certainly have to pass through many major developmental markers either to be a full replacement or to be an organ source. But why would my clone not be me, a me with whom I can do what I want, assuming that, because it is my body, I can do with it as I please?

Personhood is not identical with the genome of either the fertilized egg or the clone. I would call this position genetic reductionism: the reduction of the person to only his or her genetic structure. That is, to identify the person with the genetic structure is to say that we are only our genes. Were this position correct, the clone would in fact be me, that is, the identical person I am.

Let me use the case of human identical twins—clones of nature—as a way into this discussion. The fertilized egg divides, and the result is two individuals with an identical genetic profile. Two things are important. First, each will be a distinct individual: Jane will not be Julie, or Julie, Jane. Second, there is no claim that sharing a genetic identity diminishes in some way the dignity of either. Each is seen as a person in her own right. And even though many studies on identical twins have shown that they share many interests and similarities, even when raised in radically different environments, no one concludes from this that Jane and Julie are one single person. The fact that the twins share a genetic identity says nothing about their personal identity, value, or dignity.

Why should it be different with clones? Or on what basis would it be different? Without going too deeply into philosophical theories of individuation, it seems relatively self-evident that though the genetic profile is identical in both biological twins and clones, they are two different individuals. Minimally, one is here, and the other is there. Each has a different position from which to see the world. Each has her relation with the world. Each is a unique individual.

The ethical conclusion I would draw from this is that such individuality is privileged because it is the manifestation of a personal presence in the world. The clone, though genetically identical to its progenitor, is nonetheless an *other* being, a new being, a new presence in the world. To reduce this being to a means is to violate the clone's personal dignity—and to violate ours as well. As a thought experiment, suppose that the clone who was bred for replacement organs objected to having her heart removed. Would such objections be disregarded? If they are respected, what does this say of the replacement scenario as a whole? Many discussions of the uses of clones proceed without any such elemental moral considerations.

Reproductive Dimensions

For so many centuries, human reproduction has occurred within a biological, personal, and familial context, and we have difficulty thinking of it otherwise. Of course, IVF, in all of its manifestations, has certainly caused us to rethink that position. But while most are comfortable with the basic concept of IVF, concerns still remain about aspects of the technical context in which it is practiced. Some fear that such a context can distance the couple from each other and from the reproductive process itself. The fear is that reproduction will become production. While that fear can be alleviated to a large degree by changes in the context of the practice of IVF, with cloning the correct metaphor may indeed be production. That is, some fear that, with cloning as the means of establishing human life, the entire process will become mechanized and commercialized. One can hardly escape the memory of the "Bokanovsy Principle" from *Brave New World*, which allowed the production of "nearly eleven thousand brothers and sisters in a hundred and fifty batches of identical twins, all within two years of the same age."[19] I would argue that humans produced via cloning are humans as we are, but given the modality of their production, for how long will we consider them as such? Will this final sundering of gestation from humans and human bodies ease a transition to thinking of ourselves as objects, albeit rather sophisticated ones, but objects of production nonetheless? My argument is not an argument supporting natural reproduction. Rather, my concern is that the very means by which we produce ourselves may in fact transform our thinking about ourselves, and the transformation will be in a mechanistic direction.

Additionally, in such a context, with the laboratory as the locus of production, we will have the ultimate separation of the child from the family. To be sure, eggs and sperm will have to be obtained from humans sometime, but we know from both IVF and surrogate motherhood that eggs, sperm, and zygotes are indifferent to their origin and destiny. While being sensitive to correct feminist claims about the patriarchal nature of the family, my question is this: will we gain or lose by shifting from the family, with all its difficulties, to a laboratory? Through our genetic identity, we are linked to a family, to a lineage, to a history, and it is through these concrete biological realities that we establish at least part of our identity that has an inescapable biological dimension. If reproduction becomes mechanized through cloning, will we be put at too great a distance from a community, from a family, and from a basis for our identity? Such a problem was recognized by the engineers in the movie *Bladerunner* who gave their genetically engineered beings a history so they would not know they were engineered. But of course, that history became the basis for an identity and an evolution to human status.

Individuality

Having raised the issues of identity, let me conclude with some observations about genetic uniqueness and individuality. I have previously noted that some claim that cloning violates individuality or the individual's right to a unique genetic identity. I argue that it is important to distinguish between genetic uniqueness and individuality and that the moral priority should be placed on individuality. While a preimplantation embryo is genetically unique in that it is a new combination of the genes from the mother and father, it has not yet become individualized because restriction has not occurred. Thus, while it contains the appropriate genetic information for that organism's development, that genetic information is not morally privileged even though it is genetically unique. This genetic profile is more correctly described as what is common to all—what I previously described as our common nature.

Additionally, the developing embryo resulting from this process of biological individuation is also most appropriately what can philosophically be called a nature or human nature: the principle of activity by which a being seeks and actualizes its own fulfillment, or essentially the reason a being acts as it does. To act according to one's nature is to seek the good of one's nature or, in our case, to act according to the genetic instructions given the individual during biological development. At this stage, this is done when the organism follows the plan given it by its genes and continues its biological development. One's developmental course is set by one's nature, and this nature is set by one's genes. And ultimately, as the organism matures, its nature will lead it to seek a variety of goods for itself—food, shelter, and a desire to protect oneself are standard examples. To seek these goods is automatic on our part, an instinctual part of our nature given to us by our genetic program.

But to act on the basis of one's genetic instructions does not mean to act only on the basis of one's genetic instructions. We also seem to have a capacity to transcend our nature through acts in which we seek, for example, the good of another, or in which we love someone for his or her own sake. This capacity is the source of true liberty and is a "freedom from nature and a freedom for values."[20] We move from an act of nature to an act of the person or an act of the individual through which one freely commits one's self to a good beyond, but not contrary to, one's own nature. It is the commitment of the self to a good for the good's own sake. It is the act of love that is so taken by the good that is loved that the individual wishes this good to be shared by others. This represents, in my judgment, the supreme moment of the coincidence of personhood and individualization, for in this act of transcendence of my nature, I achieve myself in the fullest sense.

This is the act that can be actualized only by myself, and I bear the responsibility for it. Neither the motivation nor the consequences of a true act of freedom can be attributed to or communicated to another. This act describes the core of personhood. For in so acting, we transcend our own nature and reach goods that can be experienced with others and that can become the basis of a community. We do not lose our own moral identity or responsibility; rather, we find our selves and the grounds of community.

This perspective is helpful in thinking about the relation between genetic uniqueness and individuality. A genotype, even though unique, may be multiplied through cloning or may be shared by another through twinning. Frequently, much of the phenotype may be shared as well, but what is not identical and cannot be shared is individuality. Thus, while twins or clones may be, for all practical purposes, genetically interchangeable, they are not individually interchangeable. For in each of these individuals, there has been a contraction of their common nature to form an individual who can never be replicated and whose moral acts constitute a unique moral agent. The priority again is on individuality, not genetic uniqueness.

The poet Gerard Manley Hopkins describes more eloquently this process of individuation as finding one's pitch, those personal acts of self-transcendence through which we express our deepest selves. As Hopkins phrases it in the poem "As Kingfishers Catch Fire":

> Each mortal thing does one thing and the same:
> Deals out that being indoors each one dwells;

Selves—goes its self; myself it speaks and spells,
Crying What I do is me: for that I came.[21]

And in doing this, we establish our true individuality. This, I would argue, is what can never be captured through cloning, for cloning replicates only the genetic program. It replicates our human nature, but it cannot replicate the ultimate act of the individual in which he or she both expresses and becomes one's self.

CONCLUSIONS

Having examined different issues about cloning, what might be said, by way of concluding comments, regarding public policy? It is clear that not all would find my arguments persuasive or, even if persuasive, would think they should not ground public policy on them because of the religious overtones of many of my arguments. Nonetheless, I think public policy should be proposed, and I would like to suggest some components. Such considerations are more critical than ever because of a change in climate so soon after the first successful cloning. One recent article, "On Cloning Humans, 'Never' Turns Swiftly into 'Why Not,'"[22] has highlighted this climate change. The onus of the argument seems to have shifted from justifying any uses to justifying any prohibitions.

My policy recommendations would, however, focus on justifying uses of cloning.

First, public policy needs to recognize, and this would be part of its educational component, that there are three distinct forms of cloning: gene cloning, cell cloning, and whole-organism cloning. The first two types of cloning are routine in plant and animal genetics and are not the subject of these public policy proposals. They may be subject to other policies with respect to safety or environmental impact, but they are not the focus of my comments. This policy discussion is directed at whole-organism human cloning.

Second, public policy must recognize that a critical line has been crossed with respect to the cloning of a mammal. The significance is that the DNA from an adult ewe which was previously thought to have been turned off has been reactivated and become the source for a new ewe. This is a dramatic breakthrough, and while this has been done only once and only after 277 attempts, nonetheless this is a most dramatic breakthrough. There is no guarantee this procedure will work on other mammals, much less on humans. Nonetheless, that which was previously thought to be impossible has now been done. The information is public, and the knowledge cannot be retracted. Therefore, because of its profound consequences, we need to think carefully about how to use it.

Third, the research must distinguish between attempts to clone humans and research on human preimplantation embryos. I have argued in this chapter that some types of cloning research could theoretically be justified on the human preimplantation embryo up to about two weeks after fertilization. The main justification for this is my argument that until that time, the time of the biological process of restriction, the human preimplantation embryo is not yet individualized. That is, until the process of individuation is completed on a biological level, we cannot argue that we have a single individual, and being a single individual is a necessary, though not sufficient, precondition for being a person. Such research would be conducted on an organism that has a biological and teleological unity, but is not individualized. Thus, one could argue that the research is done on human nature, not on a person. Thus, I would suggest, following my argument above, that some forms of cloning research could be permitted by this policy.

On the other hand, I would argue that public policy should prohibit attempts to go beyond this two-week period to attempt to clone human individuals. I have argued that good and valid reasons exist for not permitting the cloning of humans, the majority of which have already been discussed as reasons why we might want to clone: having replacement organs or persons, having a specialized work force, ensuring an endless supply of performers in various areas of popular culture, and the interesting possibility of liking oneself so much that only one of oneself would simply not be enough.

In addition to these arguments, we need to remember that this technology has had only one success thus far. Therefore, an enormous amount of research needs to be done on animals to ensure both the efficacy and safety of the technique. An unanswered question with Dolly, for example, is whether the fact that the cell from which she was cloned was six years old will make her life span shorter. Thus, before any serious thought can be given to human application, the basic cloning technology needs to be established. Before the debate about cloning humans, we need to validate the replicability and safety of the technology.

Such policy recommendations are general, but they would permit two critical activities to occur: (1) basic research on gene and cell cloning would not be interfered with so that research on cloning technologies in animals could continue; and (2) there would be a moratorium on applications to whole-organism human cloning while the basic technology is being established, giving us time to engage in a public debate over the wisdom of whole-organism human cloning. While some might find even this minimal policy recommendation a violation of academic freedom, we should also remember that, as a practical matter, it would be scientifically and ethically irresponsible to attempt whole-organism human cloning on the basis of one experiment that required 277 attempts before success was attained. Thus, my policy recommendation is what any good scientist should say about cloning: we need to do much more basic research before we even think about human applications.

As we move forward in this debate, I would urge all to recall that common to many of the reasons supporting cloning are arguments that are crassly utilitarian and utterly self-serving. These should make us very nervous because they very clearly reveal—and would perhaps magnify—significant class, economic, and power differences in our society. Such divisions already cause enough havoc in our society. Why multiply these through cloning? But when all is said and done, when all the philosophizing and theologizing is done, perhaps the best reason against whole-organism human cloning comes from the former governor of New York, Mario Cuomo, who revealed an extraordinary amount of wisdom in commenting: "Living with the accumulated knowledge of all your imperfections, it would be hard to want to reproduce yourself and then have the arrogance to face the God who will judge you."[23]

NOTES

1. Hall and Stillman technically did not engage in cloning but rather embryo division in which they used an electrical current to cause undifferentiated cells in an embryo to divide and make more embryos for use in assisted reproduction. The method is commonly used with cattle but had not been done with humans.
2. Ian Wilmut et al., "Viable Offspring Derived from Fetal and Adult Mammalian Cells," *Nature* 385 (1997): 810–13.
3. See, e.g., Lawrence M. Fisher, "Success in Cloning Hardly lnsures Profit," *New York Times*, February 25, 1997, D1.
4. Gina Kolata reported these comments and others like them. See Gina Kolata, "Workaday World of Stock Breeding Clones Blockbuster," *New York Times*, February 25, 1997, C1 and C8. Part of the tension is between the uses of genetic technologies to solve practical problems vs. basic research, as well as the tension between those who work with animal genes and those who work with human genes.
5. See, e.g., Caryn James, "A Warning as Science Catches up on Cloning," *New York Times*, February 26, 1997, C9.
6. One of the other interesting things about cloning, of course, is that no males are needed. The nucleus of the egg is removed, and the nucleus of another cell—from a female—is inserted. If the scientists and technicians involved are female, reproduction occurs without the need of a male.
7. For his deed, Prometheus was chained by Zeus to Mt. Caucasus. Each day an eagle came and tore at his liver, and each night the liver regenerated. This lasted for many thousands of years until he was released from this torment by Hercules.
8. In this section, I follow several ideas suggested by Allen D. Verhey. See Allen D. Verhey, "'Playing God' and Invoking a Perspective," *The Journal of Medicine and Philosophy* 20 (1995): 347–64.

9. Genesis 1:1–2:3.
10. Job 38:1–38.
11. Proverbs 8:30–31.
12. See Philip Hefner, "The Evolution of the Created Co-Creator," in *Cosmos As Creation*, edited by Ted Peters (Nashville, Tenn.: Abingdon Press, 1988): 211–34.
13. Gina Kolata, "Clinics Selling Embryos Made for 'Adoption,'" *New York Times*, November 23, 1997, Al.
14. For a more thorough development of this, see Thomas A. Shannon, "Cloning, Uniqueness, and Individuality," *Louvain Studies* 19 (1994): 283–306.
15. See, generally, Norman N. Ford, *When Did I Begin? Conception of the Human Individual In History, Philosophy, and Science* (Cambridge: Cambridge University Press, 1988).
16. Philip Elmer-Dewitt, "Cloning—Where Do We Draw the Line?" *Time*, November 8, 1993, 64 and 69.
17. Here, I am explicitly using the ethics method of John Duns Scotus. For a fuller account, see Thomas A. Shannon, "Method in Ethics: A Scotistic Contribution," *Theological Studies* 53 (1993): 272–93.
18. John Donne, "Meditation XVII," in *Devotions Upon Emergent Occasions* 107, 109 (1624).
19. Aldous Huxley, *Brave New World*, (1946), 109.
20. Allan B. Wolter, O.F.M., "Native Freedom of the Will as a Key to the Ethics of Scotus," in *The Philosophical Theology of Johns Duns Scotus*, edited by Marilyn McCord Adams (Ithica, N.Y.: Cornell University Press, 1990), 148, 152.
21. Gerard Manley Hopkins, "As Kingfishers Catch Fire," in *Poems And Prose of Gerard Manley Hopkins*, edited by W. H. Gardner (1963), 51.
22. Gina Kolata, "On Cloning Humans, 'Never' Turns Swiftly into 'Why Not,'" *New York Times*, December 2, 1997, Al.
23. Jane Gross, "Thinking Twice about Cloning: Jokes Come Easily. Worries about Consequences Soon Follow," *New York Times*, February 27, 1997, B3.

8

Human Embryonic Stem Cell Therapy

Thomas A. Shannon

The recent discussions of the use of human embryonic stem cells and the decision by U.S. president George W. Bush to use federal funds to support research on only sixty-four previously existing cell lines (though it now appears that fewer than half of these lines are established and useful for research) derived from donated embryos have provoked a variety of moral and scientific claims, as well as raised questions of health care priorities. Here I review some of the scientific issues, examine several moral claims regarding the early embryo and the topic of cooperation, and identify issues related to health care and public policy.[1]

Additionally, this essay focuses on the ethical issues of only embryonic stem cell research. While it is clear that embryonic stem cells are the most desired for scientific research because of their potential, other sources of stem cells are proving to be clinically efficacious. Adult stem cells from sources such as bone marrow and the brain, as well as from other sources such as the placenta and umbilical cord, are becoming easier to obtain and their clinical usefulness is clearly established, though cost may still be an issue. Hematopoietic stem cells, derived from bone marrow cells, are being used successfully in bone marrow transplantation with the stem cells surviving in the body. Additionally, a pure line of neural stem cells from adult mice has recently been generated.[2] This summer a colleague of mine, David Adams, together with a team of students, transformed blood cells into neurons. Such potentially useful sources of adult stem cells should not be ignored simply because of the availability of a supply of more accessible—though ethically controversial—stem cells. Continued research in this direction could also eliminate the need for embryonic stem cells and thus end this debate.

We are not at that stage, however, and thus the need to focus on the use of human embryonic stem cells.

SCIENTIFIC ISSUES

Work on human stem cells began in 1998, motivated by the understanding that these cells from the blastocyst, the four- to-five-day-old embryo, had not yet differentiated and thus potentially could become any cell in the human body. This insight drives the research that focuses on how to extract these cells from the blastocyst, isolate them, grow multiple copies of them, and then coax these cells into becoming specific tissues such as muscle, nerve, or pancreatic islet cells. These specialized cells would then be implanted and repair or replace damaged tissue. Some work on animal stem cells, particularly those of mice, has been done. These cells have developed into insulin-producing cells and others into dopamine-producing neural cells—research that is of particular interest for those with diabetes and Parkinson's disease[3]—though other non–stem cell interventions such as the development of a vaccine for Alzheimer's disease are underway.[4]

A first problem in this debate is that this research remains in its infancy. Scientists have yet to achieve a critical first stage in embryonic stem cell research: the ability consistently to isolate embryonic stem cells and then tease them into the desired type of tissue, though adult stem cells can successfully be extracted from hematopoietic cells from blood marrow. And after three years of research, scientists have now developed human embryonic stem cells into blood-making cells that then became "colonies of cells, some of which were primed to make red blood cells, some white blood cells, and some platelets, the three main types of blood cell."[5] A second stage will be to discover the correct mode of delivery of the specialized cells to the part of the body that is diseased or injured. A third stage will be to determine if such specialized cells enter the body, become part of it, and begin functioning to replace the injured cells; this is the stage of tissue engineering, different from, but obviously based on, stem cell research. Once these steps are achieved, scientists will encounter another set of challenges: discovering if in fact there are long-term benefits or whether the cells multiply as expected but do not interact and replace injured cells.

I have identified only four stages in what will be a much longer and more complex research effort, and already some are arguing that the Bush proposal does not provide enough cell lines for such necessary research.[6] Several issues from the scientific perspective need to be highlighted. First, many advocates for embryonic stem cell research imply that as soon as funding begins, human trials will start and progress will be immediate, although no one has actually promised such instant progress with embryonic cells. But neither has anyone put the debate in an accurate scientific context of how very preliminary the research actually is. Thus, in my judgment, the debate has been conducted with inflated rhetoric. Embryonic stem cell research will first have to be tested on animals for efficacy and safety; only then would human research begin.

But presently, there are two radically different debates about human research going on in the United States, neither apparently within hearing distance of the other. One element of the debate assumes human trials with stem cells will start immediately; the other element seeks to heighten the review of research on humans and develop stronger guidelines and an increased role for institutional review boards. This conversation is further strained by the lack of regulations for privately funded research vs. highly regulated federally funded research or research regulated and monitored by, for example, the Food and Drug Administration. The recent death of a research subject in a gene therapy trial at the University of Pennsylvania and the death of a volunteer in a research project at Johns Hopkins University have shown that there are serious scientific and ethical issues in how research on humans is being conducted in some cases. The pressure to rapidly produce benefits from stem cell research will raise the stakes of research ethics much higher, particularly given the amount of private capital that might flow in that direction.

The second dimension in the human stem cell debate has focused on its benefits. This phase of the debate is speculative but one seldom hears that admission. To be fair, some claims are made in the hypothetical—"might produce cures," "may work"—but such caveats disappear quickly. Embryonic stem cell therapy is at present a promissory note, a scientific hypothesis, and a claim to be established. Put bluntly, no one is sure whether in fact this research will work or even if stem cells can consistently be teased into becoming specified tissues. Even in the research cited above that succeeded in coaxing embryonic stem cells into becoming blood cells, scientists, "do not yet know the signals that tell an embryonic stem cell to turn into a blood-forming cell."[7] Only long, tedious, and rigorous lab work can answer that question.

My observations do not argue that research on embryonic stem cells should not be done. Rather, I argue that very thorough, well designed, carefully monitored, federally funded studies of stem cells derived from a variety of sources, including human embryos, could be morally justified. The critical need at present is to establish the efficacy of the therapy. This can best be done with many teams working simultaneously on different approaches. Much scientific work has to precede any clinical application of stem cell therapy and, given the current state of our knowledge, we are a long way from that point. Thus the most critical current scientific issue to resolve is efficacy.

MORAL STATUS OF THE BLASTOCYST AND COOPERATION

The ethical issues surrounding the use of embryonic stem cells are complex and difficult to explicate and debate. It is clear beyond all doubt that blastocysts, the source of the stem cells, are living, are of human origin, and possess, unless twinning occurs or a clone is generated, a unique genetic code. The first question, then, is what is the moral standing of the human blastocyst, the human embryo at the early stages of cell division, specifically up to the time of its implantation into the uterine wall?

President Bush gave a partial answer in his decision for limited public funding of this research: "I also believe human life is a sacred gift from our creator. I worry about a culture that devalues life and believe, as your president, I have an important obligation to foster and encourage respect for life in America and throughout the world."[8] Use of stem cells from already destroyed fetuses "allows us to explore the promise and potential of stem cell research without crossing a fundamental moral line by providing taxpayer funding that would sanction or encourage further destruction of human embryos that have at least the potential for life."[9] President Bush's solution prohibits researchers from obtaining stem cells from IVF clinics or generating them for research, thus upholding a key principle of the prolife movement and garnering praise from Jerry Falwell and Pat Robertson—but not from the United States Conference of Catholic Bishops. Yet in order to satisfy scientists and other proponents of stem cell research, Bush allows the use of established cell lines from already destroyed fetuses (though many scientists and proponents argued that the proposal did not go far enough).

Pope John Paul II gave another perspective on this debate in an address to President Bush on July 23, 2001, during his papal visit. The pope rearticulated his position on the use of embryos by saying: "Experience is already showing how a tragic coarsening of consciences accompanies the assault on innocent human life in the womb, leading to accommodation and acquiescence in the face of other related evils such as euthanasia, infanticide, and, most recently, proposals for the creation for research purposes of human embryos, destined to be destroyed in the process."[10] The pope also called for the United States to show the world that we can be masters and not products of technology.

In a similar, though more specific response to the Bush stem cell proposal, Bishop Joseph A. Fiorenza, president of the U.S. Conference of

Catholic Bishops said: "However, the trade-off [Bush] has announced is morally unacceptable: The federal government, for the first time in history, will support research that relies on the destruction of some defenseless human beings for the possible benefit to others. However such a decision is hedged about with qualification, it allows our nation's research enterprise to cultivate a disrespect for human life. . . . The President's policy may therefore prove to be as unworkable as it is morally wrong, ultimately serving only those whose goal is unlimited embryo research."[11]

Even given these strong papal and episcopal teachings on the moral status of the embryo, I am not persuaded that the human blastocyst is a human person in the strong sense of that term as is argued above. Such a position requires, in my judgment, a static view of biology and a collapse of the traditional Aristotelian distinction between potency and act. The claim that the "human being is to be respected and treated as a person from the moment of conception and therefore from that same moment his rights as a person must be recognized"[12] ignores the fact that fertilization is itself a process requiring at least twenty-four hours to complete. (And cloning does not require fertilization at all! When might we say, then, that a cloned human begins to exist?) While some may say that a person is present after fertilization is completed, nonetheless there are several more critical stages of development before there is a clear division between inner and outer cell mass. This stage is important because the outer cell mass develops into the support structures for the embryo while the inner cell mass (the source of the stem cells) goes on to become the embryo proper. While many argue that the blastocyst at this stage is a person in potentia and, therefore, should be treated as such, a core problem is that potency is not an act. The acorn is an oak tree in potency but is not actually an oak tree. Similarly, to say that the blastocyst is potentially a person is to say that it is not yet actually a person. Collapsing the difference between act and potency only begs the question.

Modern biology shows that embryogenesis is a process. A human person, as we understand the term, arises in part out of a complex process of biochemical development. There are stages in that process but one can recognize them fully only after they have occurred. The process is organic and evolutionary, not static. I think an appreciation of this shift in an understanding of the biology of human embryogenesis is critical for a moral evaluation of the embryo. The medieval Scholastics argued that the body had to be suitably prepared before the intellectual soul could be infused. This position, termed delayed animation, was based on their understanding of the development of the embryo[13] and was in place canonically since 1591 when "Gregory XIV restricted the penalty to abortion of the animated fetus."[14] This distinction between the formed and unformed fetus was removed in 1869 as part of a restructuring of the penalties for abortion. To no one's surprise, however, the biology the Scholastics used has been shown to be inadequate. I think the same needs to be said of some of the philosophical categories used to evaluate the moral standing of the embryo. The traditional Aristotelian categories appropriated by Aquinas and other Scholastics come from a static world view that understood, for example, that all species were created as they currently exist and stood in a particular hierarchical relation, both socially and ontologically, to each other. This view does not cohere with an evolutionary world view. The theory of evolution, for example, significantly compromises the Aristotelian concept of substance for one species evolves into another, thus critically undermining the stability of a substance.

Applying analogously the still-valid insight of the Scholastics that there must be some coherence between science and philosophy, I argue that one can identify some stages in the process of embryogenesis that point to necessary, but not sufficient, conditions for personhood—stages that identify the moral standing of the embryo at various points along the developmental continuum. Here I highlight several critical stages, not the entire process of embryogenesis.[15] The first stage is the completion of the fertilization process because then one has an organism with a full genetic complement. This organism is new, genetically distinct from the parents, and may or may not be genetically unique—depending on whether or not it is a twin or was conceived through cloning. A second state is true individuality. This occurs about two weeks into the development process and is biologically and philosophically critical because after this time the cells in this particular organism become committed to being the body parts they will become in this specific body. After the process of restriction is completed, the cells of the organism can no longer be divided and become whole other organisms (and this of course is why the stem cells are obtained before this process occurs). If the organism is divided at this stage, one will get two halves. Restriction marks the presence of true individuality, a stage more morally significant than genetic uniqueness because while the genetic code can, through cloning, be theoretically copied repeatedly, individuality is unrepeatable.[16] A third stage is the development of the primitive streak, a stage that marks a transition to an organism that now has the generally recognizable structure of an embryo. A fourth stage is the emergence of various organs and the beginnings of the spinal column and brain. A fifth stage is the integration of the entire neural system, the spinal column and the brain. What is critical at this stage is that the neural system is now a fully integrated circuit in that the fetus—now about twenty-one weeks old—can now both receive stimuli and initiate some acts on its own.

These various stages are necessary, but not sufficient, biological stages in the development of personhood as we understand it. Ultimately, one cannot be a person in the full moral sense without an integrated neural system. To say that, however, does not mean that the developing organism has no moral standing or is morally valueless until all biological stages are completed.[17] It does mean, however, that the developing organism, particularly at the very early stages of embryogenesis, does not have the moral status associated with personhood. It also means that other competing values can come into play in relation to an evaluation of possible uses of that organism.

Because the blastocyst has a value deriving from simply being alive and possessing the human genome—but not the value of personhood—an argument can be made to use this organism if donated for research purposes, though few donors have actually come forward.[18] This position could also permit generating such blastocysts for research purposes, though this should be a last resort if, for example, attempts to procure stem cells from adult sources fail. This position is also open to so-called therapeutic cloning to generate specific tissues for use in a particular patient. There are two immediate reasons not to proceed with this. First, the cloning procedure itself is not scientifically established even in animals and, therefore, work on human material is premature. Second, because the therapeutic applications of cloning are limited to the generation of specific cells to be used for a particular individual in a clinical context, this procedure excludes the wider range of possible applications than the uses of a developed stem cell line would permit. Should embryonic stem cell research go forward, the initial work should be directed to the public good, not the benefit of an individual patient. (Though therapeutic cloning is the term of choice for the generation of specific tissues, I think clinical cloning would be a much better term. The act of cloning itself is not therapeutic, whereas the uses of the developed cell lines are.)

Human embryonic stem cell research should be conducted in full awareness that the research material is derived from a living human blastocyst and that in fact we are using this human tissue as a means to an end: improved health care and possible cures. Given the less than personal status of the blastocyst, I think the initial research necessary to determine the efficacy of embryonic stem cell therapy can be justified—as one element in a larger research program to determine the efficacy of stem cells from all other sources. Additionally the moral status that the blastocyst does have argues for the highest level of integrity in the design and implementation of that research as well as the availability of such research for some sort of public scrutiny. Finally, should the efficacy of adult stem cells be validated, cell lines from this source should then be generated and be the major source of cells to develop therapeutic interventions.

A second ethical question, then, is that of cooperation, specifically using materials that have been obtained in ethically compromised circumstances. This traditional problem in moral theology has recently been subject to renewed discussion, particularly with respect to the mergers of Catholic hospital systems with those of other faiths or secular perspectives.[19] But now another critical area for discussion will be: can stem cells from already destroyed embryos be used as envisioned in the Bush proposal? The essential question is: are there circumstances under which an individual or institution may participate in or benefit from an ethically compromised action of another?[20] President Bush gave one resolution in an op-ed essay: "While it is unethical to end life in medical research, it is ethical to benefit from research where life and death decisions have already been made."[21] The analogy he used here was the use of aborted human fetal tissue to develop live chickenpox vaccine that ultimately had widespread health benefits. The debate over cooperation has a long history and probably will not be resolved by the position cited above. This is particularly clear given another analogy for such research identified by Wendy Wright, the communications director for Concerned Women for America: "We should be horrified at the prospect of participating in research on embryos who are deliberately killed for the same reason that we are horrified that gold fillings were taken from the teeth of Holocaust victims."[22] Richard Doerflinger, the USCC spokesperson on bioethics, articulated the problem this way. "His [Bush's] moral principle seems to be, if the killing has already been done, we can fund this research. But by the time the scientists come forward with the next group of cell lines, that destruction will already have been done, too. And on we go. Where is the moral limit? On what basis will the president say no? I think it is an untenable and unstable policy."[23]

One assumption in the traditional discussion of cooperation is that the act with which one cooperated was in fact a moral evil. Given that the act in question is the removal of cells from a blastocyst, that assumption may have to be revisited or at least reexamined. Given my previous analysis for the moral standing of the blastocyst, its destruction would certainly be a premoral evil, but not a moral evil. Since the blastocyst does not have the moral standing of full personhood, its destruction is killing but not murder for there is no person who can be the subject of such a moral wrong. The scientists who accept such cells for research are not, in this analysis, cooperating in a morally evil act.

A second question is whether a patient could make use of vaccines derived from embryonic stem cells. Several responses can be given. First, the patient need not intend the destruction of the embryos and thus any cooperation would not be formal. Second, the moral distance between the use of the vaccine by the patient and the original research is so great as to render any cooperation remote at best. Finally, the original act must itself be immoral and that at least can be questioned as noted immediately above.

Perhaps there is a better moral question: is the science of embryonic stem cell therapy sufficiently established to justify the killing of the blastocyst for research purposes? I think a cautious answer of yes can be given, but only insofar as the research that should go forward is well controlled and publicly monitored. Pilot programs, perhaps federally funded, should be used as kind of pilot programs to gain critical information so that the next stage of research will be on scientifically sound grounds.

These positions discussed above, however, do not in themselves justify proceeding with stem cell research. What I have argued thus far is a possible justification of the use of embryonic stem cells in research. This argument does not resolve several other issues that I think are at least equally, if not more, important for the stem cell debate.

PUBLIC POLICY ISSUES

I have argued that an ethical case can be made for using human embryonic stem cells in research. In this section, I wish to step back and identify two other issues: If the research goes forward, how should it be conducted; second, should it go forward?

First, if the research goes forward—and indeed it will—I argue that such research, including the development of new cell lines from embryonic tissues from IVF clinics, should be conducted so that the research process will have at least some minimal monitoring. For example, such research could be subject to review by an institutional review board and be subject to the traditional peer review process associated with grant reviews. Or there could be scrutiny by a federal agency such as the Food and Drug Administration or perhaps an analogue to the National Institutes of Health Recombinant DNA Advisory Committee. The national debate revealed that private labs are conducting cloning research on human embryos as well as generating embryos for research. It is in the public interest to have such research conducted with public review or under the direction of a licensing board such as the United Kingdom[24] has established. Even though investors are currently avoiding biotech stocks,[25] market incentives are present and may intensify with the consequence that those concerns rather than ethical standards may drive the research. For example, when human trials with specialized stem cells begin, who will monitor these trials? Some biotech companies such as Geron in California and ACT in Massachusetts do have ethics committees to review their scientific research. Will these same committees review their animal experiments and human studies? Because the resolution of President Bush provides federal funds only for the sixty-four stem cell lines already in existence, embryonic stem cell research in the United States may continue unmonitored in private labs. There is not a group of "mad scientists" creating monsters in private labs—though the three scientists who are insisting on their right to engage in reproductive cloning does demonstrate the presence of extremists even among scientists. Market forces, the fiscal incentives for patents, licensing fees and market shares, as well as fame, fortune, and prestige will shape how research is done. A publicly monitored and regulated (and perhaps funded) process, such as a licensing board as in the British model, a variant of the Investigational New Drug Application from the Food and Drug Administration, or perhaps an analogue to the National Institutes of Health Recombinant DNA Advisory Committee might be more beneficial in the long run for the integrity of the research as well as patient well-being.

There are many other complicated issues such as intellectual property rights, licensing fees, and royalties that have yet to be resolved. For example, the resolution of the problem proposed by President Bush apparently gives a head start on the research to the private labs that have already developed stem cell lines with tissue from aborted fetuses or donated embryos. But such may not be the case. The Wisconsin Alumni Research Foundation at the University of Wisconsin holds Patent 6,200,806. This patent, "which covers both the method of isolating the cells and the cells themselves, gives the Wisconsin foundation control over who may work in the United States with stem cells, and for what purpose."[26] Previous licensing arrangements of the foundation with Geron Corporation are complicating the use of stem cells by other researchers, although some progress has been made for their use in basic research.[27] How this patent will control U.S. embryonic stem cell research, even if the stem cells are obtained from the offshore labs, remains unresolved—as well as any international implications this patent may have.

But there is a second question that is seldom asked: should this research be done at all? This serious question raises the issue of what kind of health care system is appropriate in the United States as well as a question of social justice—the distribution of the benefits of research.

Research initiatives such as stem cell therapy continue to replicate the dominant trend of high-tech, high-cost rescue medicine that has driven the U.S. health care system for the past several decades. The focus will continue to be cure rather than prevention. In part this focus has dominated because of the great success in curing many diseases and those successes, combined with better diets, improved sanitation, and clean drinking water, helped more and more to live longer and longer. And since no good deed goes unpunished, many now experience a whole new range of diseases that are proving to be exceptionally difficult to cure. The war against cancer has been going on for at least three decades, that on AIDS for at least two decades, and we are beginning to invest more money in curing very complex diseases such as Alzheimer's. The cures are no closer, though significant progress has occurred, particularly with certain forms of cancer. But the fact remains that curing is an attempt to reverse the harm done by the disease, a process much more complex than attempts to prevent the disease initially.

Should health care continue to go in this direction? The question should at least be debated. Present U.S. health care policy, particularly the allocation of research dollars, seems to be driven by lobbying groups or moral extortion. The diseases with the best celebrity spokespersons—who themselves may or may not be afflicted with the disease—testify before congressional committees. Parents show up pleading for research funding so their children, who are also at the hearings, will not die. The strong suggestion is that Congress will be morally accountable for the fate of the sick if they do not provide funding. A concomitant problem, of course, is who is able to testify before such committees? One does not see in congressional hearing rooms the poor, the marginally insured, or the uninsured—all of whom suffer because they cannot obtain decent health care, or perhaps even decent housing or diets, factors that also have a profound impact on one's health.

It is clear beyond all doubt that adequate funding cannot be provided to seek cures for all diseases. How then might U.S. citizens go about debating how this country's resources are to be allocated? I am not arguing against research into the cure of diseases, but I am arguing that the current funding mechanisms that are essentially driven by the group that has the best lobby are totally inadequate and unjust.

Perhaps the most critical issue that needs debating is not the ethics of embryonic stem cell research but the very system of high-tech health care that is predominant in the United States. Perhaps it is time to think of an emphasis on prevention rather than cure or on the availability of minimal health care services for all citizens. Such an emphasis would benefit many more people through its focus on prenatal care, well baby care, vaccinations, proper nutrition, avoiding of smoking and excess drinking (though the recent study by Phillip Morris for Czechoslovakia did demonstrate long-term financial benefits to the state, such as fewer payments for long-term health care for the elderly, of not discouraging smoking, a practice that leads to early death[28]). Poverty and poorly constructed and maintained housing, for example, are major contributors to diseases such as asthma. Funding decreases in programs such as WIC (Women, Infants, and Children—a program offering nutritional supplements to those individuals) resulted in higher long-term health care payments.

Prevention is a hard sell. It makes for boring TV. Who wants to watch a physician counseling a person to stop smoking when he or she could be watching the fast-paced medical drama *ER* with all its exciting high-tech interventions (even though the success rates of such interventions are inflated)?[29] Who wants to add an exercise routine to an already over-scheduled day? Who wants to be moderate all the time? And increased smoking rates, drug use, and drinking among the young certainly indicate a very difficult health care future.

Yet I would still argue that the most critical issue that needs to be faced is not the ethics of embryonic stem cell research or any of the other questions raised in light of the human genome project. These questions are important and must be confronted. But the more critical issue is access to health care and insurance. If stem cell therapy proves successful but one's insurance plan does not cover such treatment or if one has no insurance to begin with, what good is this research? How will it benefit the millions of uninsured or underinsured citizens? The U.S. health care system has developed as a fee-for-service business and is in a transition to various forms of managed care. That the market will not resolve the current health care crisis is shown by the failure of many health maintenance organizations to remain viable while providing minimal, let alone, adequate care. Trying to resolve the health insurance problems through private employment works to some degree for the employed—but what about all the rest of the citizenry?

The embryonic stem cell debate can provide a moment of pause in which to reflect on health care in the United States. Should current health care policies continue? Clearly, this direction will bring benefits to many, but I suspect many, many more will not receive those future benefits because they are currently not receiving many of the benefits the health care system already provides. Perhaps, then, the more critical question to be debated is not the moral standing of the blastocyst and its use in research but the justice of the current U.S. health care system.

NOTES

1. For a review of many of the issues discussed in this essay, see Suzanne Holland, Karen Lebacqz, and Laurie Zoloth, eds., *The Human Embryonic Stem Cell Debate: Science, Ethics, and Public Policy* (Boston: MIT Press, 2001).

2. Rodney L. Rietze et al., "Purification of a Pluripotent Neural Stem Cell from the Adult Mouse," *Nature* 412 (16 August 2001): 736–39. Although the work was done on mice, the authors claim, "This demonstrates that a predominant, functional type of stem cell exists in the periventricular region of the adult brain with the intrinsic ability to generate neural and non-neural cells" (p. 736).

3. Sheryl Gay Stolberg, "A Science in Its Infancy, but with Great Expectations for Its Adolescence," *New York Times*, 10 August 2000, A17.

4. For a report on the development of this vaccine, generated from a mouse genetically engineered to develop Alzheimer's disease, see D. Schenk, D. Adams et al., "Immunization with Amyloid-ß Attenuates Alzheimer's Disease-like Pathology in the PDAPP Mouse," *Nature* 400 (8 July 1999): 173–77. This vaccine is now in phase II of human research trials.

5. Nicholas Wade, "Team Says It Coaxed Human Cells to Produce Blood," *New York Times*, 4 September 2001, A17.

6. Editorial, "Downside of the Stem Cell Policy," *New York Times*, 31 August 2001, A18.

7. Wade, "Team Says It Coaxed Human Stem Cells to Produce Blood," ibid.

8. "Excerpts from Bush Address on U.S. Financing of Embryonic Stem Cell Research," *New York Times*, 10 August 2001, A16.

9. Ibid.

10. "Pope John Paul II Addresses President Bush," www.americancatholic.org/News/StemCell (accessed 23 April 2003).

11. "Catholic Bishops Criticize Bush Policy on Embryo Research," www.nccbuscc.org/comm/archives/20001/o1-142.html.

12. *Donum vitae*, I, 1. The citation can also be found in Thomas A. Shannon and Lisa S. Cahill, *Religion and Artificial Reproduction* (New York: Crossroad, 1988), 149. The Instruction is careful to note that the Catholic Church has not taken a philosophical position on the time of ensoulment. However, "From the moment of conception, the life of every human being is to be respected in an absolute way" (Shannon and Cahill, Introduction, 4, ibid., 147).

13. Sources for this discussion are John T. Noonan Jr., *Contraception: A History of Its Treatment by the Catholic Theologians and Canonists* (Cambridge: Mass.: The Belnap Press of Harvard University Press, 1965) and John Connery, S.J., *Abortion: The Development of the Roman Catholic Perspective* (Chicago: Loyola University Press, 1977).

14. See Connery, ibid., 212.

15. For an excellent presentation of the embryogenesis process, see www.zygote.swarthmore.edu.

16. For a further development of this point see Thomas A. Shannon and Allan B. Wolter, O.F.M., "Reflections on the Moral Status of the Pre-embryo," *Theological Studies* 51 (December 1990): 603–26; Thomas A. Shannon, "Cloning, Uniqueness and Individuality," *Louvain Studies* 19 (1994): 283–306; and Thomas A. Shannon, "Human Cloning: Religious and Ethical Issues," *Valparaiso University Law Review* 32 (1998): 773–92.

17. Implicit recognition of this moral standing is given by the fact that not many people want to be involved in the actual destruction of the frozen embryos. See Gina Kolata, "The Job Nobody at the Fertility Clinic Wants," *New York Times*, 26 August 2001, A20.

18. Already questions are being raised about the availability of such embryos for research as well as the integrity of the sixty-four cell lines approved by President Bush. See Gina Kolata, "Researchers Say Embryos in Labs Are Not Available," *New York Times*, 26 August 2001, A1 and A20.

19. For discussions of this see M. Cathleen Kaveny and James F. Keenan, S.J., "Ethical Issues in Health-Care Restructuring," *Theological Studies* 56 (1995): 136–51; and James F. Keenan, S.J., "Prophylactics, Toleration and Cooperation: Contemporary Problems and Traditional Principles," *International Philosophical Quarterly* 29 (1989): 205–20.

20. For a discussion of this with respect to the stem cell controversy, see "Fact Sheet: Embryonic Stem Cell Research and Vaccines Using Fetal Tissue," United

States Conference of Catholic Bishops, Pro-Life Activities, www.nccbuscc.org/prolife/issues/bioethics/vaccfac2.html.

21. George W. Bush, "Stem Cell Science and the Preservation of Life," *New York Times*, 12 August 2001, Week in Review, 13.

22. Laurie Goodstein, "Abortion Foes Split over Plan on Stem Cells," *New York Times*, 12 August, A22.

23. Ibid.

24. For a brief review of the United Kingdom model, see Nicholas Wade, "Stem Cell Studies Advance in Britain," *New York Times*, 14 August 2001, A1 and A14.

25. Andrew Polack, "The Promise in Selling Stem Cells," *New York Times*, 26 August 2001, Sec. 3, 1 and 11.

26. Sheryl Gay Stolberg, "Patent on Stem Cell Studies Puts U.S. Officials in Bind," *New York Times*, 17 August 2001, A1.

27. "Agreement of an accord with the Wisconsin Alumni Research Foundation permitting the use of such cells was recently reported. The agreement applies only to government-employed scientists, and covers only basic research; if scientists want to use the cells as therapies, they will have to renegotiate." Sheryl Gay Stolberg, "Many Approved Stem-Cell Lines Aren't Ready to Study, U. S. Says," *New York Times*, 6 September 2001, A20.

28. For the full text of this enlightening analysis, see www.americanlegacy.org/Czech.

29. For an analysis of these claims with regard to cardio-pulmonary resuscitation, see David L. Wheeler, "A Sociologist in the ER Punctures the Myth of CPR," *Chronicle of Higher Education*, 15 October 1999, A21. The articles notes: "Although one study found that cardio pulmonary resuscitation works about 75 per cent of the time on television, a variety of regional studies suggest it works at least less than 3 percent of the time in real life."

9

Ethical Issues in Stem Cell Therapy: From the Micro to the Macro

Thomas A. Shannon

As with many developments in bioethics in the last decade, stem cell research and the promises implicit in it hit both the scientific community and the media with a cosmic explosion, leaving in its dust many people—scientists and ethicists, among others—who were both astonished and confused about this latest development. What I want to highlight in this presentation is that this therapy has both micro and macro issues, and while many have focused on the micro issues—the status of the organism from which the stem cells are obtained—the macro issues—commitment to high-tech medicine and therapies that are directed to the privileged—are often neglected. I will argue that while there are ethical justifications possible for obtaining stem cells from human embryonic tissue, nonetheless the larger social issues argue for at least caution in pursuing stem cell research.

STEM CELL RESEARCH: THE MICRO ETHICAL ISSUE

The most critical micro ethical issue in stem cell research is the source of the stem cells themselves: the human embryo. These have been obtained in two different ways: one is from germ cells from aborted fetuses and the other is from cells from embryos not used in *in vitro* fertilization (IVF).

In the former case, the particular ethical issue is that of cooperation in the evil of abortion—assuming of course that abortion is a moral wrong.

If abortion is not a moral wrong, the particular micro ethical problem is ensuring the separation of the consent to use the tissue for this purpose from the consent to the abortion to ensure that the abortion is not coerced. If one thinks that abortion is wrong, one could still argue that researchers using the tissue are at a sufficient moral distance from the abortion to be able to use the tissue.

The latter case—using embryos from IVF clinics or, as some would want, generating embryos to obtain their stem cells—presents the more difficult, for many, ethical problems. If one's position is that personhood begins with the process of fertilization, then one would argue that no human embryo could be used in this way. I wish to develop the other possibility: namely, that while bearing a unique—at least thus far—genetic code and while assuredly human, the embryo at this stage is not a person and thus some interventions can be done.

For me this argument has three levels. First, though genetically unique, the cells in the very early embryo—at the zygote and blastomere stages—are totipotent. That is, they are not yet differentiated or committed to the particular cells they will become in the body—heart cells, liver cells, etc. These totipotent cells have the capacity to become any body part, hence their obvious desirability for stem cell research. However, the very totipotency of the cells, while conferring some biological unity on the developing organism, also strongly suggests the absence of a more critical ontological level of organization.

Second, again—though genetically unique at this stage and because the cells are yet totipotent—the developing organism is not individualized. That is, while this organism has a biological unity and organization, its cells can still be separated through twinning or divided through embryo division and thus different whole organisms obtained. The blastomere is an organism that can be divided into parts each of which can become another organism. Such an organism is divisible such that its parts can become wholes. And such an organism is by definition not an individual. For an individual is literally indivisible—or if divided, is divided in such a way that what remains are parts only. The individual is no longer there.

Third, though the NBAC describes this as a human life form, I think a more suitable, though perhaps more complicated, way of thinking of the early embryo till the time of differentiation is that of a biological expression of human nature. I base this on the philosophy of individuation of the medieval philosopher Duns Scotus who I assure you had never heard of stem cells, much less anything else in modern genetics. But I think elements in his theory of individuation lend themselves in a particularly helpful way to evaluate morally the status of the blastomere. The term Scotus uses is common nature and is a part of his larger theory of individualization. This common nature is essentially the basis for the definition of an entity—what all members of a particular class share in common. But what is important for Scotus is that this common nature is indifferent to being either a particular individual or referring to all members of that particular class. Thus, for Scotus, the common nature needs something else—an individualizing principle—to make it a particular being of this class. Additionally this common nature has a unity to it, but a unity less than a numerical unity. That is, the common nature is not an individual being—which would by definition give it a numerical unity—but rather has a unity characteristic of or common to members of the entities it defines. Scotus's principle of individuation constricts, as he says, the form of this common nature into an individual, rendering this being individual, unique, and distinct from all others of the same species. This process of individuation also renders it indivisible, thus giving it a numerical unity and, therefore, making it incapable of being divided into two wholes.

We can think of the blastomere as the biological equivalent of Scotus's concept of common nature because while this entity is genetically distinct from its parents, it is not yet individualized. This does not occur until after the process of restriction is completed, some two weeks after fertilization is completed. To my mind this process is an interesting biological complement to Scotus's concept of common nature's being constricted into an individual. After this process is completed, the cells become committed to being specific cells in specific body parts. This is the biological beginning of true individuality and marks a critical ethical line.[1]

Until the line of individuation is crossed biologically, these cells are indifferent to becoming specific cells in this particular body by virtue of their totipotency; they are not morally privileged by virtue of individuality or, a fortiori, by personhood. They are morally privileged by being human cells, cells that manifest the human genome, and an entity that represents the essence of human nature. Essentially such research would be utilizing cells that in fact are the reality of human nature in its most basic form and meaning. Such a presentation of human nature in the blastomere is preindividual and prepersonal. And because this is human nature and not individualized human nature (the minimal definition of personhood), I argue that cells from this entity could be used in research to obtain stem cells. Clearly consent from those from whom such entities come must be obtained for this research and the blastomeres must be handled with respect. But ultimately, such research is not research on a human person; it is research on human nature and in principle is morally permissible.

Yet a word of caution needs to be added here. To use such cells in research is to objectify human nature, to make it a means to an end. While it is clear that—all things being equal—it is ethical to do research on humans and while it is clear that humans can donate body parts for research, it is another thing to generate human embryos exclusively for research.

While I would not argue that ending the life of such an organism at the totipotent stage is murder—for there is no subject of such an act—such a means of obtaining stem cells does reduce the embryo to an object. Therefore, we need to be exceptionally cautious about such use and perhaps make the use of such cells the last resort.

STEM CELL RESEARCH: THE MACRO ETHICAL ISSUES

The promise of stem cell research is significant and important. There is the likelihood of using such cells for drug development, toxicity testing, study of developmental processes, learning about gene control, and developing specific cells for use with bone marrow, nerve cells, heart muscle cells, and pancreatic islet cells. The further promise of such research is captured in the common description of such cells as immortal. The hope is that such cells can be directed to develop in specific ways so they can be directly injected into the problem area, and these cells can replace or compensate for the diseased cells there.

Several comments are in order. First, the promise. One of the characteristics of research into genetics is exaggeration and inflation of claims or, as it is called by some, gene hype. One of the most extreme examples was the claim by a senior scientist at the beginning of the genome project that the success of the project would lead to solving the problems of poverty and homelessness. While everyone can appreciate the ridiculousness of this claim, other claims are not seen as exaggerations or as significant promissory notes. What we have yet to recognize is the immense and substantive gap between discovery and cure. This is not an argument against stem cell research per se. It is a call to recognize inflated claims that are used to justify commitment of money to a process that is highly experimental and untested. The claim is not the reality, but one would not always know that from listening to discussions of various discoveries.

Second, commitment to stem cell research is a commitment to business as usual in the medical community. That is, a commitment to stem cell research is a commitment to high-tech, very expensive rescue medicine. Now clearly, that is the dominant mode of medicine practiced across much of the United States, particularly in the wealthier areas. And clearly high-tech medicine is where the money is to be made. Pursuing stem cell research continues this practice and continues to draw large sums of money from other possible uses. As with all other research efforts, particularly in the area of genetics, stem cell research offers great promise for the cure of diseases. But the success of such research will be extremely costly, and the product of such research will also be costly because investors will be seeking an adequate return on their investment.

Additionally, as E. Richard Gould points out, a commitment to secure property values in human body parts such as embryonic tissue commits us implicitly to specific health policies. First, we will seek out cures for diseases and turn "away from discovering the underlying social and environmental causes of diseases." Second, we would commit ourselves to a health policy "that holds that health status is improved by access to newer and better treatments." Finally, this policy would suggest that "disease ought to be viewed as an individual problem, specifically as a problem of the individual's genetic code, instead of as a social problem."[2]

The very difficult social question is: is this the way to continue to go with research and medicine? Should we continue down the track of high-tech, rescue medicine with its emphasis on intervention and cure or is it time to have a substantive conversation on other models of medical practice and medical intervention? I do not want to pick on stem cell research, but it is a clear example of another promissory note in modern medicine with the payment to be picked up at the cost of other interventions and research into human well-being, as well as the delivery of health care itself.

Third, who will be the beneficiaries of stem cell research? The rhetoric is that all will benefit. But in the meantime, the benefit will be reaped by two groups: (1) those who are insured and whose insurance will cover the treatment, and (2) those who can afford to buy it. Because of the millions of Americans who are uninsured, underinsured, or whose insurance will not cover such experimental protocols, the vast majority of citizens will not have access to whatever benefits come from this therapy. Additionally, depending on how the science goes, researchers may focus on single-cell genetic diseases because these are easier to identify and target. But again this narrows the field of application considerably. While it may be the case that research on single-cell genetic diseases using material derived from stem cells provides the possibility of cure for many who might otherwise be without a remedy for their disease, nonetheless this is still a significant directing of the scarce resource of research money to a small population. Thus, even should the therapy prove successful, the number of people who stand to benefit from it are a small subset of the whole population and perhaps even a small subset of all those with genetic diseases.

Fourth, and already alluded to, is the cost of such treatment in both experimental and therapeutic stages. This kind of research is time consuming and labor intensive. While computers and other automated systems aid tremendously, the main part of the work is both theoretical—understanding genetic structures and planning the research—and practical—carrying out the experiments and studying their results. Well-equipped labs with sophisticated equipment in addition to a highly trained staff are the basic entry requirements for this kind of work. And if much of the research will be funded by private capital because of current federal difficulties over the use of human embryos, one can be sure investors will want a return. And as mentioned, that return will take the form of expensive therapy. Patients whose incomes are not in the upper 5 percent would not be able to pay for such therapy—as they are not able to pay for many other therapies in our current medical system. Even those who have good insurance plans will have difficulties because of the continued restriction on what will be covered by such plans and a growing reluctance to fund experimental therapies. Again the number of possible beneficiaries narrows.

In an early paper on justice and the Human Genome Project (HGP), Karen Lebacqz suggested that one way to achieve justice would be through some form of price controls in any medications or interventions resulting from HGP research since this research is supported in part by public funding.[3] Thus, while private capital investments are an important source of funding for the HGP, significant monies are derived from, for example, the National Institutes of Health (NIH). My point is not that such funding is wrong or improper; rather, it is to suggest that there is an obligation in justice to acknowledge these public sources of funding and a relatively easy way would be to follow Lebacqz' suggestion of some form of price controls.

CONCLUSIONS

I would argue that the micro ethical debate over the use of early human embryos is not the key factor in resolving the larger stem cell debate. While I think it is the case that an argument can be made for the use of such cells, another more critical variable is the consequences of such an objectification of human nature in this way. Thus, while I would argue that in principle there is an argument for the use of such cells, the consequences of such use might be more problematic than we currently realize.

However, I think the more important argument is what I have identified as the macro argument, the social context in which such cells will be used. Here I would argue that at a minimum we should be very cautious about going down the path of stem cell research. First, what we have with stem cell research is yet another promissory note from scientists. Let us at least develop some more specific understanding of the therapeutic implications through animal research on stem cells. And if one opposes such research on animals, I would note that a fortiori that one should also oppose it on humans for exactly the same reasons. Second, developing research on stem cells commits us to the same medical model that is already causing such a complex of problems in health care. Business as usual is not going to resolve our health care crisis. Perhaps it is time to apply the brakes in some areas to try to solve some problems in other areas, i.e., public health. Finally, those who are most ill and most vulnerable will most likely not have access to the benefits of therapies derived from stem cell research, should they in fact materialize. Insurance plans will probably not cover such experimental treatments and those without insurance or those who are underinsured will not have the funds available to purchase such services. Is focusing on developing expensive cures for a narrow range of diseases the most effective use of public money and social resources?

I am not opposed to the HGP or research deriving from it. I am not in principle opposed to stem cell research. What I am suggesting is a moratorium—first, to develop some experimental results to see what we are getting into and, second, to force us to think through what kinds of health care reforms we need and how stem cell research might fit into that, if at all.

NOTES

1. For an expanded development of this idea, see Thomas A. Shannon, "Cloning, Uniqueness, and Individuality," *Louvain Studies* 19 (Winter, 1994): 283–306.
2. E. Richard Gould, *Body Parts: Property Rights and the Ownership of Human Biological Material* (Washington, D.C.: Georgetown University Press, 1996), 37.
3. Karen Lebacqz, "Fair Shares: Is the Genome Project Just?" in *Genetics: Issues of Social Justice*, edited by Ted Peters (Cleveland: The Pilgrim Press, 1998), 97.

10

The Bioengineering of Planet Earth: Some Scientific, Moral, and Theological Considerations

James J. Walter

As the twentieth century has come to a close and the new millennium has dawned, we have witnessed spectacular developments in biotechnology and molecular genetics. It is becoming rare that we can pick up a daily newspaper or listen to a national news program and not be presented with one more scientific or medical discovery that stretches our moral imaginations. Many of these discoveries have enabled us on a global level to genetically engineer plants and animals. Now, with the announcement in June 2000 that the Human Genome Project (HGP) has mapped and sequenced over 90 percent of our genetic code, we will soon have the ability to genetically engineer ourselves. (In February 2001, an initial analysis of the data from the HGP indicated that humans have far fewer genes than originally believed: rather than having 100,000 to 140,000 genes, we have somewhere between 30,000 and 40,000 and maybe as few as 26,000 genes. This is only twice as many as a fruit fly or a roundworm!)[1] For some, these technologies are welcomed as the means to advance human purposes and goals around the globe; for others, these developments are viewed with great caution and suspicion.

Humans have used living organisms in agriculture and food production since early history and have derived important pharmaceuticals from them for decades. Controversies over biotechnology began to arise after the development in the mid-1970s of novel and powerful techniques that allow dramatically increased control over the design of living organisms. Recombinant DNA technology, which allows specific pieces of genetic information to be moved between species, has raised most of the ethical and religious concerns.

In the last twenty years, scientists have discovered that DNA is virtually interchangeable among animals, plants, bacteria, and humans. Thus, I will use the phrase "bioengineering" to include any of the following: (1) the attempt to modify the DNA of any plant, animal, or human or (2) any attempt to transfer the DNA from one organism or species to another, e.g., the transfer of a segment of human DNA to an animal. In the latter case, the modified species is called transgenic.

This chapter will develop in three parts. In the first part, I will describe in turn the major global developments in the genetic engineering of plants, animals, and humans. Under each, I will discuss briefly the ethical issues involved. In the second part, I will focus on two theological issues from the Christian tradition that are at stake in the area of genetic engineering and biotechnology. Finally, the conclusion will focus on constructing an ethical agenda on these issues as we move further into the twenty-first century.

THE GENETIC ENGINEERING OF PLANTS, ANIMALS, AND HUMANS

Plants

Most of the genetic engineering projects that are in progress globally fall under one of the following three types: (1) engineering for improved crop production and quality, e.g., herbicide or pest resistance; (2) engineering for improved health, e.g., edible vaccines for the prevention of hepatitis B or the building of medicines into cornflakes; and (3) biopharming, or the engineering of plants for alternative nonfood uses, e.g., rather than building expensive factories we will be able simply to grow the chemicals needed for making plastics, detergents, and construction materials. In the latter case, in the next ten years it is possible that 10 percent of the U.S. corn acres will be devoted to this type of bioengineering.[2] Some of these projects involve the insertion of animal genes into plants to create what are called transgenic plants. For example, the DNA Plant Technology Corporation in New Jersey (USA) added a gene from the Arctic flounder to make a tomato frost resistant.[3] In addition, corn and tobacco plants have been engineered to accept human genes as part of their DNA in order to make drugs to fight cancer and osteoporosis.[4]

There have been mixed reactions to these biotechnologies around the world. In the United States, where people tend to be pragmatic and more willing to take risks, it seems that the biotech companies are moving forward with great speed to produce these transgenic plants.[5] Things may be moving forward even more rapidly in China where in 1986 Chinese scientists were already aggressively pushing for governmental efforts to bioengineer plants to feed their 1.2 billion people.[6] On the other hand, in Europe the situation has been much different. People in the United Kingdom[7] and Germany have staged many protests against the introduction of these genetically altered foods into their countries,[8] but the Swiss, after many protests, voted decisively in June 1998 to reject a proposal to outlaw the production and patenting of genetically modified plants, ostensibly because they did not want to surrender Swiss leadership in biotechnology.[9]

There are several important ethical issues here, and I will organize them into the categories of *benefits* and *burdens* that are connected to the bioengineering of plants. First the *benefits*. With the growing world population, especially in China and India, this technology could repeat the fantastic abundance of grain production that occurred in the "Green Revolution."[10] Scientists promise other benefits as well, e.g., more diverse and improved foods, foods that are more fit for their environment, crops that are less reliant on the use of chemicals, and finally, transgenic plants that can become "biological factories" to produce drugs like interferon for humans.[11]

Notwithstanding these great potential benefits, there are several important *ethical burdens* associated with these technologies. First, the expansion of plant bioengineering by large multinational companies could create an unfair competition for small, privately-owned farmers who desire to grow crops on organic farms. Second, there are many environmental and safety issues that need to be addressed. There could be long-term risks to humans associated with the eating of these transgenic plants.[12] These plants likewise could present environmental risks to surrounding crops, e.g., by spreading new viruses.[13] This very scenario seems to have occurred in Iowa in November 2000 when the genetically-engineered Star-Link corn that was planted by one farmer spread into and destroyed the entire crop of another farmer's nonengineered corn. In addition, the insertion of foreign genes into the DNA of these plants could cause allergies to humans, since these new genes might produce the proteins that cause allergies. Finally, there is the *ethical issue of fair allocation of public funds* to develop these engineered crops. Some argue that it is simply unjust to fund these bioengineering efforts and not fund at an equal level other efforts to develop nonchemical methods of pest control in an environmentally sound system of agriculture.[14]

Animals

Humans have been selectively breeding animals for centuries, but now we have the capacity to genetically engineer these animals for human benefit in a very short time span. Briefly, in this section I will discuss two forms of animal genetic engineering, along with the issue of patenting the results of these efforts. Finally, I will point out some of the ethical issues.

The first form of genetic engineering is concerned with modifying the genetic makeup of an animal for human benefit. For example, scientists at Harvard Medical School in the United States created a new life form, the *oncomouse*. This mouse had been genetically engineered to develop cancer so it could be used in the design of new drugs for the treatment of human diseases.[15] Another instance of this form of engineering is the creation by GenPharm International of the *knockout mouse*, which had a particular gene eliminated from, or knocked out, of all its cells so the effects of the elimination could be studied to benefit humans.[16]

The second form of genetic engineering is the creation of transgenic animals by using recombinant DNA (rDNA) modification. In one instance, scientists at Monsanto have altered the DNA of bacteria so that they can produce a cow hormone (bovine somatotropin, or BST) that increases milk output.[17] In January 2001 scientists announced that they had bred a type of catfish whose DNA has been laced with DNA from salmon, carp, and zebrafish. Many are voicing fears that this new catfish will escape and wipe out other fish species, as well as the plants and animals that depend on those fish to survive.[18] In January of this year, researchers in Oregon were able to genetically engineer a monkey, one of the closest relatives of humans—a major advance toward producing primates with human genetic diseases as a research tool to find new cures.[19] A fluorescent gene from a jellyfish was inserted into the monkey so that the presence of the gene could be detected. In other instances, scientists have been creating transgenic clones of animals by using the cloning technology developed by Ian Wilmut in Scotland when he created the sheep Dolly and Polly. Some of these animals, like Polly, carry human genes that have been inserted into their DNA and are cloned in order that they might produce human proteins to be used for new medicines, cancer treatments, and even to create a new no-calorie sweetener used in cooking.[20] In December 2000, scientists in Massachusetts seemed to have entered a type of *Jurassic Park* by taking a somatic cell from a mountain goat (a *bucardo*), which had just died and which was the last of its species. The somatic cell was placed into an enucleated egg from another species, and thus Celia, as the extinct goat was called, is now cloned as the revival of an extinct species.[21] In another instance, research scientists have inserted human genes into the DNA of animals, and then other genes in their DNA are knocked out so that, when these animals' organs are harvested for human transplantation, the human immune system will recognize the tissue of the transplanted organ as human. Therefore, the lack of available human organs for transplantation will soon be offset by harvesting animal organs. Each time the DNA of these animals is modified or transgenically altered, the labs performing these experiments want to patent not only the procedures used but also the modified genes and the pharmaceuticals developed from the experiments.

I will discuss only two types of ethical issues connected to the bioengineering of animals. The first relates to the *proposed benefits and the potential risks* of this technology. Frequently, the proposed *benefits* are the following: (1) genetically engineered animals could increase productivity of food and fiber; (2) the development of such animals could boost economic growth; and (3) such animals could, in theory, model a variety of human diseases and conditions, and the harvesting of their organs can benefit humans.[22] On the other hand, there are many potential *risks* associated with this type of bioengineering. We need to be concerned with unanticipated consequences that may affect the changed animal. We need to be concerned with the narrowing of the gene pool of these animals and with modifying the pathogens that inhabit the animal hosts. Finally, we need to be concerned with the patenting of genetically engineered animals and the socioeconomic risks this poses world wide.[23]

The second set of ethical issues is related to the *use of animals in research* that is aimed at benefiting not them but humans. The pain and suffering inflicted on them in the experiments can be great and must be weighed against the real benefits derived from the experiments.[24] Of course, some argue that these animals have rights and thus should not be used in any experiment that does not directly or indirectly benefit them. And many question the ethics of patenting the transgenic life forms created in these laboratories.

Humans

Perhaps the most contentious and morally problematic area in genetic engineering concerns those technologies that have the capacity to alter or duplicate (through cloning) the genetic code of humans. I will address briefly four forms of genetic manipulation of the human genome and then mention several of the ethical issues connected to them.

First, there are recombinant DNA (rDNA) technologies, in which the genetic material from one organism or species—in this case human genes—is recombined with the genetic material of another organism. There are good indications that we will continue to use this technology, and even *splice* foreign genes from other species into humans in the near future.

Second, the completion of the HGP will present us with tremendous possibilities over the next twenty to thirty years. Clinical medicine will be revolutionized because physician scientists will be able to modify our genetic code that causes the disease rather than treating the symptoms of the disease.[25] The results from the HGP will enable us to modify our DNA at either the somatic cell level or at the germ-line cell level (egg and sperm) in order to cure genetic defects or to prevent disease. Likewise, however, we will be able to use this same technology to alter the DNA in either of these types of cells—somatic or germ-line—in order to enhance or engineer genes to produce a "better" human being,[26] whatever that might mean! Multiply the effort to create a "better" human being at the global level, and we have some very important ethical issues to confront.

Third, we have learned recently that an American (Dr. Zavos at the University of Kentucky) and an Italian scientist (Dr. Antinori in Rome) have announced their plans to clone a human being in an undisclosed Mediterranean country within the next eighteen months to two years.[27] The procedure is called somatic cell nuclear transplant technology, where the nucleus of a body cell from one individual is transplanted into and fused with an egg whose own nucleus has been destroyed. It is commonplace today to clone all types of animals in this fashion. The desire, at least in the United States, to provide women with the newest assisted reproductive technologies whereby they might be able to become pregnant with a child genetically linked to them, will be difficult to resist. Human cloning through this procedure also has the possibility of duplicating oneself, of replacing a deceased child, or of providing an endless supply of organs for transplantation to oneself.

Finally, there are the most recent attempts to use pluripotent stem cells from early embryos (ES cells) or from aborted fetuses (EG cells) to generate new cells or organs for patients.[28] Because these cells are capable of giving rise to almost every type of cell or tissue in the body, scientists can harvest these cells from either of the two sources and then direct the development of their cell lines into whatever organ is needed for transplant. If one further combines this technology with somatic cell nuclear transplant cloning, which was approved in Britain in January 2001, then the pluripotent stem cell lines from the clone could be coaxed into creating a new organ for transplant without the possibility of tissue rejection. In this case, the embryonic clone would be destroyed at the blastocyst stage (approximately four to six days after fertilization) in order to retrieve the inner cell mass that contains the pluripotent stem cells.

There are many *ethical issues* related to these genetic technologies with humans. I will only have the time to mention a few of the more important ones.

Though there are obviously great *benefits* that could be derived from some of these genetic technologies, there are many *risks* as well. How safe and effective will the technologies be?[29] In addition, there are ethical problems with *people's privacy* over their genetic code, since the gathering of genetic information could easily lead to discrimination in a medical setting or in the workplace.[30] Third, there are a number of ethical problems with the *patenting of human genes* and with claims to ownership over genetic material.[31] Fourth, what is the *moral status of the preimplantation embryo*? Because in many of these genetic technologies, e.g., pluripotent stem cell research and cloning, experiments are performed on these entities in which they are destroyed in order to harvest cells within them. Finally, at the macro level of individual societies and of the global community, there are many important *issues of social justice*. Who will have access to these high-tech and very expensive technologies, and at what cost are we willing as a human community to develop them for the use of a wealthy few?

THEOLOGICAL CONSIDERATIONS

In this part of my presentation I would like to address two important theological issues from the standpoint of the Christian religious tradition. Theological or religious perspectives can raise questions about who we are as humans created by God and about what kind of authority we have over creation. They can also raise questions about whether there are limits to our responsibility to alter God's creation. My goal here is show that the *moral* judgments at which Christian believers arrive on the genetic engineering of plants, animals, and humans are often informed and guided by convictions and beliefs of a *religious* nature. In other words, I want to suggest that the moral decisions that Christians come to concerning whether or not to support bioengineering depend partially on a religious context of meaning. Thus, in the case at hand, theological convictions can provide perspectives on and engender attitudes toward genetic engineering.

Image of God: Should We Have Control over the Genetic Future of Plants, Animals, and Humans?

Christians believe that all humanity is created in the image and likeness of God (Gen. 1:26–27). However, the great Christian tradition has used at least two different interpretations of how humans stand in that image, and these diverse models almost inevitably lead to different moral evaluations about genetic engineering and the control we ought to have over God's creation.

The first interpretation defines humanity as a steward over creation. Our moral responsibility is primarily to protect and to conserve what the divine has created and ordered. Stewardship is exercised by carefully respecting the limits placed by God in the orders of biological nature and society.[32] If we are only stewards over both creation and the genetic heritage of God's creatures, then our moral responsibilities do not include the alteration of what the divine has created and ordered through nature. Our principal moral duties are to remain faithful to God's original creative will and to respect the laws that are both inherent in creation and function as limits to human intervention. In this view, most, if not all, forms of bioengineering of plants, animals, and especially humans would be morally suspect, if not prohibited. The second interpretation of the *imago dei* defines humans as cocreators or participants[33] with God in the continual unfolding of the processes and patterns of creation. As created cocreators, i.e., as beings who do not create *ex nihilo* as God does,[34] we are both utterly dependent on God for our very existence and simultaneously responsible for creating the course of human history. Though we are not God's equals in the act of creating, we do play a significant role in bringing creation and history to their completion.[35] Proponents of this interpretation would be more likely to support morally plant and animal engineering, and they would likely approve morally of some therapeutic interventions to modify the human genome.

Playing God

There is a second theological framework that has shaped the discussion of these topics in genetics and bioengineering. It is concerned with the question of whether or not humans are exceeding their limits, and thus "playing God," by intervening into the very material that constitutes all life. This framework implies two theological themes: (1) the status of human DNA, and (2) the sovereignty of God and the divine ownership of creation.

When the U.S. President's Commission on Genetic Engineering submitted its report in 1982, it noted that there was an objection from religious groups that scientists were "playing God" in their recombinant DNA (rDNA) research.[36] Part of this concern was over the status of human DNA: is it sacred and thus beyond the boundaries of human manipulation and control, or is it more or less equal in status to other matter and thus open to human intervention and control?[37] If the former, then scientists are improperly playing God when they bioengineer DNA because they overstep the boundaries given to them by God; if the latter, scientists are properly playing God, i.e., serving God's own purposes, when they intervene to produce some benefit for humans.

Some have argued theologically that human DNA and the genome itself are sacred because they possess characteristics integral to human identity and personhood. Furthermore, DNA provides the biological blueprint for humans created in the image of God, and it is even possible to accord to this genetic material the social and cultural functions of the soul.[38] In addition, some theologians have argued that the very image of God in humans pervades human life in all its parts, and this certainly includes DNA.[39] Several have argued also that to patent human genes is to play God improperly because such actions take away God's sovereign ownership over these genetic materials. On the other hand, most theologians have argued that, though human DNA is a cause of great wonder, it does not possess a sacred status but is like other biological matter. Thus in principle it may be altered within certain moral limits.[40] In this view, one could argue that theists believe only God is sacred, and thus everything else is God's creation. This argument claims that there is no significant difference between DNA and other complex chemicals, and so there is no distinctly religious ground for objecting to the patenting of DNA.[41]

CONCLUSION: FUTURE AGENDA

Bioengineering and biotechnology will continue to pose enormous challenges in all the areas under consideration, and in fact the pace will only hasten as each of these areas develops and applies its new-found knowledge to human and nonhuman life. Thus, a clear ethical agenda is necessary to confront these challenges. I will restrict my remarks to only the area of the human. The list is not sacrosanct, but many would probably agree that there are about seven areas that need to be included in the agenda. There is no necessary rank ordering of these priorities, but surely some are more important than others, or the urgency of one will take priority over others.

First, there will be the need to put increasing pressure on the scientific community and biotechnology companies to protect the dignity of their research subjects. There is more and more evidence that harms are accruing to human subjects in various research protocols. The asthma experiment in July 2001 at Johns Hopkins University—where Ellen Roche, a 24-year-old healthy volunteer, was killed—is but one example. Part of this agenda must also deal with the protection of early embryos in research protocols and with their creation and use to benefit the health of others. Second, we will have to introduce new ethical and legal measures to protect the privacy of genetic information. We will need to sort through who should have rightful access to this knowledge and who should be excluded. Since genes are inherited within families, do family members also have a right to this knowledge that could adversely affect their health? Genetic discrimination is a real possibility, especially in the workplace or in the clinic, but it is also a possibility in terms of securing health and life insurance.

Third, now that the human genome is basically mapped and sequenced and the scientific community is moving forward with not only identifying defective genes but also with the mapping of the proteins that cause disease, what will we do with this knowledge? Since our knowledge will outstrip our clinical abilities to cure many of these genetic diseases, how will we use this knowledge when counseling patients? Will we simply create the life-time identity of a sick person when we tell an adolescent that she will get adult-onset Alzheimer's disease when she's eighty years old? Fourth, genetic screening of targeted populations must come under severe scrutiny. The chances for discrimination are enormous here, particularly if this type of screening is simply routine at birth or during a visit to the clinic.

Fifth, human gene transfer, either with somatic cells or germ-line cells, for purposes of therapy or enhancement, will probably be introduced on a regular basis within the next twenty or thirty years at the clinical level. This technology could be combined with human cloning techniques and embryonic stem cell research, and thus different combinations of genetic medicine will develop over the next few decades. The agenda must sort out the differences between therapy and enhancement, and it must decide whether we should ever cross into the germ line with any of these technologies.

Sixth, there are huge marketplace issues at stake, especially with the multinational pharmaceutical companies that stand to gain enormous profits. The fair pricing of the new drugs and therapies that will be developed must be constantly assessed. Ownership of the gene sequences, whether human or nonhuman, must be decided at both the ethical and legal levels. Finally, and very importantly, we must ensure that the development and use of these biotechnologies will not simply benefit the few who are rich enough to afford them. We will need to focus our attention on the equity of the delivery of health care in a way that we have not in the past. Not only are there millions in the United States who virtually go without any health care, there are millions and millions throughout the world who basically have no access to the wonders of modern medicine. Just social policies must be forged, not only in the scientific and medicine-rich West, but around the world.

NOTES

1. Aaron Zitner, "Humans Need Fewer Genes than Thought to Survive," *Los Angeles Times*, February 11, 2001, A1 and A44.

2. Aaron Zitner, "Fields of Gene Factories," *Los Angeles Times*, June 4, 2001, A1 and A7, at A7.

3. Calgene Inc., a California (USA) biotechnology company, was the first to bioengineer a tomato (the MacGregor tomato) by slowing down the tomato's natural softening process in order that the fruit could ripen a few days longer on the vine instead of being picked green and hard for easy shipping (Stuart Gannes, "Take a Tomato, Add a Gene," *Self*, March 1993, 152–55, at 152; and Pamela Weintraub, "The Coming of the High-Tech Harvest," *Audubon*, July–August 1992, 92–103, at 94). Also, Thomas A. Shannon, *Made in Whose Image? Genetic Engineering and Christian Ethics* [New York: Humanity Books, 2000], 9).

4. Ronald Kotulak, "A Brave, New World Emerging at 'Biopharms,'" *Chicago Tribune*, February 8, 1998, 1 and 12, at 12.

5. Michael Specter, "Europe, Bucking Trend in U.S., Blocks Genetically Altered Food," *New York Times*, July 20, 1998, A1 and A8, at A1.

6. Ted Plafker, "First Biotech Safety Rules Don't Deter Chinese Efforts," *Science* 266 (November 1994): 966–67, at 966.

7. Warren Hoge, "Britons Skirmish over Genetically Modified Crops," *New York Times International*, August 23, 1999, A3.

8. Nigel Williams, "Agricultural Biotech Faces Backlash in Europe," *Science* 281 (August 7, 1998): 768–71, at 768.

9. Irene Harnischberg, "Genetics Initiative Defeated," 2, http://www.mercurycenter.com:80/premium/world/docs/swiss08.html, June 9, 1998.

10. Charles C. Mann, "Crop Scientists Seek a New Revolution," *Science* 283 (January 15, 1999): 310–14, at 310.

11. Shannon, 13.

12. Shannon, 13.

13. Warren E. Leary, "Gene Inserted in Crop Plant Is Shown to Spread to Wild," *New York Times*, March 7, 1996, B14; and Keith Schneider, "Study Finds Risk in Making Plants Viral Resistant," *New York Times*, March 11, 1994, A16.

14. For a further discussion of the genetic engineering of plants, see Reinhard von Broock, "Chancen und Risiken von Gentechnologien an Pflanzen," 16–22; Sanjay K. Mishra "Environmental aspects of genetic engineering—Challenges and prospects," 52–56; and Ren-Zong Qui, "Social and political impact of genetic engineering," pp. 81–91 in *Genes the world over: Die Bewertung von Gentechnologie an Pflanzen und Tieren in der Sicht verschiedener Kulturen*, edited by Sybille Fritsch-Oppermann (Loccum: Evangelische Akademie Loccum, 1998). Also, Rebecca Goldburg et al., *Biotechnology's Bitter Harvest: Herbicide-Tolerant Crops and the Threat to Sustainable Agriculture*, March 1990, 7.

15. Sarah Franklin, "Life," in *Encyclopedia of Bioethics*, rev. ed., vol. 3, edited by Warren Thomas Reich (New York: Simon and Schuster Macmillan, 1995), 1345–52, at 1350.

16. Christopher Anderson, "Researchers Win Decision on Knockout Mouse Pricing," *Science* 260 (April 2, 1993): 23–24, at 23.

17. Barnaby J. Feder, "Monsanto Has Its Wonder Hormone, Can It Sell It?" *New York Times*, March 12, 1995, 8.

18. Aaron Zitner, "Gene-Altered Catfish Raise Environmental, Legal Issues," *Los Angeles Times*, January 2, 2001, A1 and A8.

19. Thomas H. Maugh II, "Healthy Monkey Is Genetically Engineered by Oregon Researchers," *Los Angeles Times*, January 12, 2001, A1 and A28.

20. Kotulak, 12.

21. The first successful attempt at cloning an animal and using another species' enucleated egg and womb to gestate the animal was an Asian gaur (named Noah), which died in January 2001 two weeks after birth from an ordinary disease, dysentery. Also, Aaron Zitner, "Cloned Goat Would Revive Extinct Line," *Los Angeles Times*, December 24, 2000, A1 and A18.

22. Bernard E. Rollin, "Genetic Engineering: Animals and Plants," in *Encyclopedia of Bioethics*, rev. ed., vol. 2, edited by Warren Thomas Reich (New York: Simon and Schuster Macmillan, 1995), 932–36, at 934.

23. Rollin, 933–34.

24. Marie Buy, "Transgenic Animals, Animal Welfare and Ethics," http://www.acs.ucalgary.ca/~browder/guidelines.html, October 28, 1997.

25. For example, a recent report using human twins has indicated that genetic factors seem to account for at least 30 percent of all cancers (Mestel 2000, 1). Physicians could develop genetic therapies specifically aimed at the individual genetic code of patients to cure their cancer rather than using the type of broad-spectrum chemotherapy available today.

26. Maurice A. M. de Wachter, "Ethical Aspects of Human Germ-Line Gene Therapy," *Bioethics* 7 (April 1993): 166–77, at 166.

27. Aaron Zitner, "'Genie out of the Bottle' on Human Cloning," *Los Angeles Times*, January 28, 2001, A1 and A24.

28. National Institutes of Health, "Stem Cells: A Primer," http://www.nih.gov/news/stemcell/primer.html, September 2002 (accessed 23 April 2003).

29. In September 1999 Jesse Gelsinger was injected with engineered genes by the University of Pennsylvania researchers to cure the boy's rare liver disease, but he died as a result of the procedure. His death tragically illustrates that some of these experiments have not yet been proven safe to use on humans.

30. See Thomas A. Shannon, "Ethical Issues in Genetics," *Theological Studies* 60 (March 1999): 111–23.

31. See M. Cathleen Kaveny, "Jurisprudence and Genetics," *Theological Studies* 60 (March 1999): 135–47.

32. Thomas A. Shannon, *What Are They Saying about Genetic Engineering?* (New York: Paulist Press, 1985), 21.

33. James Gustafson, the influential Protestant theologian, has preferred to describe our role in creation as participants rather than as cocreators. He argues that the divine continues to order creation, and we can gain some insight into God's purposes by discovering these ordering processes in nature. (James M. Gustafson, *Ethics from a Theocentric Perspective—Volume Two: Ethics and Theology* [Chicago: University of Chicago Press, 1984], 294.)

34. Philip Hefner, "The Evolution of the Created Co-Creator" in *Cosmos as Creation: Theology and Science in Consonance*, edited by Ted Peters (Nashville: Abingdon, 1989), 211–33.

35. Ted Peters, "'Playing God' and Germline Intervention," *The Journal of Medicine and Philosophy* 20 (1995): 365–86, 377–79; and Ann Lammers and Ted Peters, "Genethics: Implications of the Human Genome Project," in *Moral Issues and Christian Response*, edited by Paul T. Jersild and Dale A. Johnson (New York: Harcourt, Brace and Jovanovich College Publishers, 1993), 300–06, at 302.

36. President's Commission for the Study of Ethical Problems in Medicine and Biomedical and Behavioral Research. *Splicing Life* (Washington, D.C.: Government Printing Office, 1982), 54.

37. For a discussion of some of these issues, see Bernard Baertschi, "Devons-Nous Respecter Le Génome Humain?" *Revue de Théologie et de Philosophie* 123 (1991): 411–34.

38. Mark J. Hanson, "Religious Voices in Biotechnology: The Case of Gene Patenting," *The Hastings Center Report* 27 (November–December 1997): 1–21, at 4.

39. Richard D. Land and C. Ben Mitchell, "Patenting Life: No," *First Things* 63 (May 1996): 20–23.

Rosie Mestel, "Study Ties Most Cancer to Lifestyle, Not Genetics," *Los Angeles Times*, July 13, 2000, A1 and A28, at 21.

40. For example, see The Catholic Bishops' Joint Committee on Bioethical Issues 1996, 32. Jeremy Rifkin, who has protested against this view, is a notable exception. See his *Algeny* (New York: Penguin Books, 1983).

41. Ronald Cole-Turner, "Religion and Gene Patenting," *Science* 270 (October 6, 1995) 52.

Permissions

Chapter 1
James J. Walter, "Theological Issues in Genetics," *Theological Studies* 60 (March 1999): 124–34.

Chapter 2
Thomas A. Shannon, "Ethical Issues in Genetics," *Theological Studies* 60 (1999): 111–23.

Chapter 3
James J. Walter, "Catholic Reflections on the Human Genome," *The National Catholic Bioethics Quarterly* 3 (Summer 2003): 275–83.

Chapter 4
Thomas A. Shannon, with Allan B. Wolter, O.F.M., "Reflections on the Moral Status of the Pre-embryo," *Theological Studies*, 51 (December 1990): 603–26.

Chapter 6
James J. Walter, "Human Gene Transfer: Some Theological Contributions to the Ethical Debate," *The Linacre Quarterly: A Journal of the Philosophy and Ethics of Medical Practice* 68 (November 2001): 314–34.

Chapter 7
Thomas A. Shannon, "Human Cloning: Religious and Ethical Issues," *Valparaiso Law Review* 32 (Spring 1998): 73–92.

Chapter 8
Thomas A. Shannon, "Human Embryonic Stem Cell Therapy," *Theological Studies* 62 (December 2001): 811–24.

Chapter 9
Thomas A. Shannon, "From the Micro to the Macro," in *The Human Embryonic Stem Cell Debate: Science, Ethics, and Public Policy*, ed. Suzanne Holland, Karen Lebacqz, and Laurie Zoloth (Cambridge, Mass.: The MIT Press, 2001), 177–84.

Chapter 10
James J. Walter, "The Bioengineering of Planet Earth: Some Scientific, Moral and Theological Considerations," *New Theology Review* 15 (September 2002): 41–54.

Index

About the Authors

Thomas A. Shannon is professor of religion and bioethics in the Department of Humanities and Arts at Worcester Polytechnic Institute. He is the author or editor of more than twenty-five books and the author of more than forty articles about ethics and bioethics. His most recent works are *Made in Whose Image? Christian Ethics and Genetic Engineering* and *Catholic Perspectives on Peace and War*. He has been a board member of both the Society of Christian Ethics and the Catholic Theological Society of America.

James J. Walter is Austin and Ann O'Malley Professor of Bioethics and is director of the Bioethics Institute at Loyola Marymount University in Los Angeles. He has published several books and is a regular contributor to many theological and ethics journals in the field of bioethics. Together with Thomas A. Shannon and Timothy E. O'Connell, he recently edited *A Call to Fidelity: On the Moral Theology of Charles E. Curran.*